應用英文寫作系列之六

Writing Effective Marketing Promotional Materials
有效撰寫行銷英文

By Ted Knoy
柯泰德

Illustrated by Ming-Jay Chen
插圖：陳銘杰

Ted Knoy is also the author of the following books in the Chinese Technical Writers' Series（科技英文寫作系列叢書）*and the Chinese Professional Writers' Series*（應用英文寫作系列叢書）：

An English Style Approach for Chinese Technical Writers
《精通科技論文寫作》

English Oral Presentations for Chinese Technical Writers
《做好英文會議簡報》

A Correspondence Manual for Chinese Technical Writers
英文信函參考手冊

A Correspondence Manual for Chinese Writers
《英文信函參考手冊》

An Editing Workbook for Chinese Technical Writers
《科技英文編修訓練手冊》

Advanced Copyediting Practice for Chinese Technical Writers
《科技英文編修訓練手冊進階篇》

Writing Effective Study Plans
《有效撰寫讀書計劃》

Writing Effective Work Proposals
《有效撰寫英文工作提案》

Writing Effective Employment Application Statements
《有效撰寫求職英文自傳》

Writing Effective Career Statements
《有效撰寫英文職涯經歷》

Effectively Communicating Online
《有效撰寫英文電子郵件》？

This book is dedicated to my wife, Hwang Li Wen.

序　言

　　科技對世界的影響無遠弗屆，隨著資訊、通訊與交通的便捷，國際間的關係越來越密切。在此宏觀的架構下，台灣產業除了憑藉既有之優秀製造技術能力，也需發展研發設計（ODM）能量及朝向積極發展自創品牌（OBM），透過國際行銷通路建立實力，立足全球。

　　國際行銷要能針對每個階段的使用者，用不同的行銷包裝與訴求、定價、通路、客戶服務等，將產品讓消費大眾接受，若再能搭配優良的品牌形象，將可為台灣企業經營帶來更大的利潤與附加價值，促進台灣產業在全球分工體系與全球市場中取得競爭優勢，台灣趨勢科技的Trend及明基電通的BenQ品牌，即是朝這個方向努力發展的例子。

　　生動活潑的行銷文字與圖像的結合，除了讓消費者對您的產品或服務留下深刻印象外，往往也是激發消費者購買的意念；而流利與專業的英文運用與溝通能力，更是國際行銷與品牌推動關鍵之一。

　　柯泰德先生（Ted Knoy）以他深厚的英文造詣，近年來寫過多本科技英文寫作叢書，包含科技英文寫作系列及應用英文寫作系列，如：精通科技論文寫作、科技英文編修訓練手冊……等，使科技人在鑽研科技的同時，也能將科技研發的成果向國際展現。而這本《有效撰寫行銷英文》書，則是特別針對行銷相關英文，加以解說並輔以範例，加深讀者之印象，並以六個視覺化的情境，訓練讀者的口說、聽力、閱讀及寫作能力，是從事國際行銷、管理工作者值得參閱的書籍。

工業技術研究院　院長

李鍾熙

i

Table of Contents

Foreword

Professional writing is essential to the international recognition of Taiwan's commercial and technological achievements. The Chinese Professional Writers' Series seeks to provide a sound English writing curriculum and, on a more practical level, to provide Chinese speaking professionals with valuable reference guides. The series supports professional writers in the following areas:

Writing style

The books seek to transform old ways of writing into a more active and direct writing style that better conveys an author's main ideas.

Structure

The series addresses the organization and content of reports and other common forms of writing.

Quality

Inevitably, writers prepare reports to meet the expectations of editors and referees/reviewers, as well as to satisfy the requirements of journals. The books in this series are prepared with these specific needs in mind.

Writing Effective Marketing Promotional Materials is the sixth book in The Chinese Professional Writers' Series.

"Writing Effective Marketing Promotional Materials" (《有效撰寫行銷英文》)為「應用英文寫作系列」（The Chinese Professional Writers' Series）之第六本書，本書中練習題部分主要是幫助國人糾正常犯寫作錯誤，由反覆練習中，進而熟能生巧，提升有關行銷英文的寫作能力。

　　「應用英文寫作系列」將針對以下內容逐步協助國人解決在英文寫作上所遭遇之各項問題：

A.寫作型式：把往昔通常習於抄襲的寫作方法轉換成更積極主動的寫作方式，俾使讀者所欲表達的主題意念更加清楚。更進一步糾正國人寫作口語習慣。

B.方法型式：指出國內寫作者從事英文寫作或英文翻譯時常遇到的文法問題。

C.內容結構：將寫作的內容以下面的方式結構化：目標、一般動機、個人動機。並了解不同的目的和動機可以影響報告的結構，由此，獲得最適當的報告內容。

D.內容品質：以編輯、審查委員的要求來寫作此一系列之書籍，以滿足讀者的英文要求。

This writing workbook aims to instruct students on how to compile effective marketing promotional materials. First, how to forecast market trends is introduced, including how to briefly state the rationale for developing this product, point out the market trends in the area of development while describing the financial aspect of developing this product, visually depict the organization's strategy and approach to developing the product and summarize market survey results, particularly in terms of the product's commercial potential and categories of product application. How to describe how a product or service is developed is then introduced, including how to introduce the current status of product or service development, describe its market value, point out unique features and characteristics of product or service development, list major manufacturers of this product or service in Taiwan and explain the rationale for further product or service development in this area. Next, how to describe a project for developing a product or service is outlined, including how to provide the rationale for taking on the project, present market data that justifies the project's feasibility, spell out the immediate and long term goals of the project, describe the project' s distinguishing features, point out the strategy employed to successfully complete it and conclude the presentation by pointing out the anticipated merits of the project and the positive impact that it will have. Additionally, how to introduce a company or organization is described, including how to briefly overview the industry to which one's organization belongs, describe one' s organization's mission as well as highlight its historical development, introduce the organizational structure, highlight recent technical accomplishments within the organization and point towards the organization's future directions. Moreover, how to introduce a division or department in a company or organization is introduced, including how to describe the setting of one's division or department within the larger organization, highlight the organizational structure of the division or department, point out the staff's strengths and educational backgrounds/ work experience in a particular field, spell out the missions of the division or department, elaborate on the manufacturing or research capabilities within the division or department and list the technical services (e.g. industrial and consultancy) that the department or division offers. Furthermore, how to introduce a technology is described, including how to briefly explain the factors (internal and external) factors that influence development of this technology in Taiwan, point out the unique characteristics of this technology development in Taiwan, list the objectives of how to further develop this technology, define the role of this technology in relation to environmental, manufacturing or technology problems, and list applications of this technology made so far, highlighting any particular characteristics or features that are unique to Taiwan's circumstances. Finally, how to introduce an industry is described, including how to briefly highlight the general characteristics of this industry in Taiwan, point out the difficulties encountered in industrial development, describe current activities of the particular industrial sector, elaborate on the available technologies that are employed by the industry, discuss the related research and development facilities in Taiwan and how they assist/collaborate with that particular industry.

Each unit begins and ends with three visually represented situations that provide essential information to help students to write a specific part of a career statement. Additional oral practice, listening comprehension, reading comprehension and writing activities, relating to those three situations, help students to understand how the visual representation relates to the ultimate goal of writing an effective career statement. An Answer Key makes this book ideal for classroom use. For instance, to test a student's listening comprehension, a teacher can first read the text that describes the situations for a particular unit. Either individually or in small groups, students can work through the exercises to produce concise and well-structured marketing promotional materials.

簡 介

　　本書主要教導讀者如何建構良好的行銷英文。書中內容包括：

1. 預測市場趨勢：

　　開發此產品的原因，科技層面：(1)相關產品；(2)潛在利潤。財務層面：開發領域的市場趨勢。

　　產品開發的組織策略和方案：(1)策略的界定（包括最高目標及主要重點）；(2)執行策略方案的界定。

　　市場調查：(1)產品的商業潛力；(2)產品運用的範疇。

2. 產品或服務研發：產品開發的現階段狀況；市場的評價；產品開發的獨創性和特色；台灣同類產品的主要製造廠商；在既有領域下未來產品開發的原因。

3. 專案描述：專案的背景；市場的訊息；專案的目標；專案的重要特色；專案的策略；專案的動向。

4. 公司或組織介紹：簡要陳述組織所屬之產業的概況；組織的使命；組織的發展沿革；組織的架構；組織最新的科技成就；結論 (未來發展方向)。

5. 組或部門介紹：介紹部門所屬的組織或公司；部門的組織架構；部門的人才來源和教育背景；部門的使命；部門之製造或研究的能力；部門提供的產業服務。

6. 科技介紹：影響台灣科技發展的因素；科技發展的特色；發展科技的目標；界定科技的地位：扼要解釋在環境、製造或科技方面的問題；目前從事的應用科技：(1)應用科技的特色或特徵，(2)特殊的個案；科技市場的契機；繼續從事科技應用的未來挑戰。

7. 工業介紹：此種產業在台灣的一般特色；產業所面臨的困境；簡述一或二個目前活動的重點；產業採用的科技；台灣相關的研究發展設備。

　　書中的每個單元呈現六個視覺化的情境，經由以全民英語檢定為標準而設計的口說訓練、聽力、閱讀及寫作四種不同功能來強化英文能力。此外，本書也非常適合在課堂上使用，教師可以先描述單元情境，讓學生藉由書中練習，循序在短期內完成。

Unit One

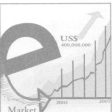

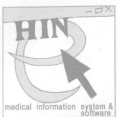
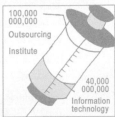

Forecasting Market Trends

預測市場趨勢

1. Briefly state the rationale for developing this product.
   開發此產品的原因，科技層面：(1)相關產品；(2)潛在利潤

2. Point out the market trends in the area of development while describing the financial aspect of developing this product.
   財務層面：開發領域的市場趨勢

3. Visually depict the organization's strategy and approach to developing the product.
   產品開發的組織策略和方案：(1)策略的界定（包括最高目標及主要重點）；(2)執行策略方案的界定

4. Summarize market survey results, particularly in terms of the product's commercial potential and categories of product application.
   市場調查：(1)產品的商業潛力；(2)產品運用的範疇

Vocabulary and related expressions

learning medium	學習媒介
time constraints	時間限制
space limitations	空間限制
personal learning environments	個人化的學習環境
temporal and spatial constraints	時間及空間限制
digital content applications	數位內容的運用
multidisciplinary professionals	有關各種學問的專家
asynchronous learning	非同步遠距教學
vocational training	職業訓練
medical-oriented information	以醫學為方向的資訊
clinical practice	臨床的看診
annual expenditures	每年的開支
outsourcing	向國外採購零配件／工作外包
hardware maintenance	硬體維護
profound impact	深深的影響
incentives for investment	投資的動機
sustained economic growth	持續的經濟成長
raising awareness of	引起察覺
daunting regulations	使人怯步的管制
marks a milestone	建立一個里程碑
prompt transactions	及時的交易
validation	批准／確認
traffic congestion	交通擁塞
increasing incidence of	日漸增加的影響範圍
popularization of ...	的大眾化
nutritionally balanced diet	滋養地平衡飲食
successful product commercialization	成功的產品商品化
chemotherapy medicine	化學療法用藥
continuous technological advances	持續的科技進長
digitalized consumer products	數位商品
highly promising area	具有高度發展前途的區域
increased commercial interest	日增的商業利益
ethnic appeal	種族（上）的吸引力
catering to individual preferences	投合個別的偏愛

Situation 1

Situation 2

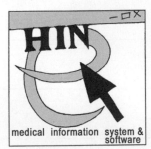 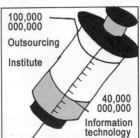 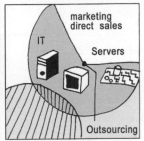

Situation 3

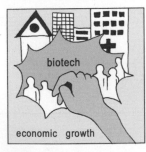 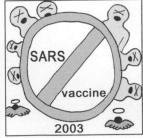 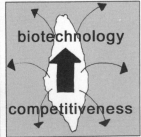

A Write down the key points of the situations on the preceding page, while the instructor reads aloud the script from the Answer Key. Alternatively, students can listen to the script online at www.chineseowl.idv. tw

Situation 1

_____

_____

_____

Situation 2

_____

_____

_____

Situation 3

_____

_____

_____

B Oral practice I

Based on the three situations in this unit, write three questions beginning with **What**, and answer them. The questions do not need to come directly from these situations.

Examples

What does e-learning eliminate?

Time constraints and space limitations faced by classroom instruction

What is a notable example of personal learning environments constructed by e-learning?

Internet-based English learning websites

1. _____

_____

2. _____

_____

3. _____

_____

C Based on the three situations in this unit, write three questions beginning with **When**, and answer them. The questions do not need to come directly from these situations.

Examples

When was the National Health Insurance scheme established?

In 1995

When did computerization of hospital operations increase to 57%?

In 1996

1. _____

   _____

2. _____

   _____

3. _____

   _____

D Based on the three situations in this unit, write three questions beginning with **Why**, and answer them. The questions do not need to come directly from these situations.

Examples

Why has the biotech sector emerged as one of the main areas of industrial growth in the new century?

Owing to its profound impact on human quality of life

Why has the biotech industry become a potential engine of sustained economic growth?

Because the Taiwanese government has offered numerous incentives for investment in and development of this industry

1. _____

    _____

2. _____

    _____

3. _____

    _____

E Write questions that match the answers provided.

1. _____

_____

Taiwan's inability to effectively prevent epidemics and treat infected patients

2. _____

_____

Strengthen its research capabilities, develop patented technologies and attract biotechnology professionals with expertise in multidisciplinary fields

3. _____

_____

Owing to its profound impact on human quality of life

F Listening Comprehension I

Situation 1

1. What has e-learning in Taiwan recently emerged as a highly promising learning medium for?

   A. digital content applications

   B. enhancing traditional classroom instruction

   C. constructing personal learning environments

2. What do Internet-based English learning websites commonly adopt?

   A. the sharable content object reference model (SCORM)

   B. professionals in art design, information technology, marketing and curriculum design

   C. personal learning environments

3. How much in revenues did the e-learning market in the United States generate in 2003?

   A. US$ 500,000,000

   B. US$ 300,000,000

   C. US$ 400,000,000

4. In what area does e-learning represent a highly promising area?

   A. Internet-based English learning websites

   B. the SCORM standard

   C. digital content applications

5. How are learners able to enhance their competitiveness in school and at work?

   A. through interactive websites

   B. through the Greater China market

   C. through potential revenues

Situation 2

1. When was the National Health Insurance scheme established?

    A. in 1996

    B. in 1997

    C. in 1995

2. In what area has medical-oriented information technology been extensively adopted?

    A. in clinical practice

    B. in outsourcing organizations

    C. desktop systems

3. How much in revenues did the US information industry outsource to the medical services sector in 2001?

    A. $100 billion dollars

    B. over $US 2,809,000,000

    C. over $US 1,755,000,000

4. What can be problematic?

    A. integrating future applications into existing systems

    B. computerizing hospital operations

    C. training new employees

5. What continues to be the largest area for outsourcing?

    A. applications development

    B. re-engineering

    C. information technology

Situation 3

1. How has the Taiwanese government prioritized the biotech sector in national development strategies?

 A. by focusing on the pharmaceutical industry

 B. by emphasizing the need to promote immunity in the community

 C. by offering numerous incentives for investment

2. What heightened concerns over Taiwan's inability to effectively prevent epidemics and treat infected patients?

 A. the lack of research and development capabilities

 B. the SARS crisis in 2003

 C. incomplete clinical research and product trials

3. Why is the local biotechnology market scale small?

 A. the inability to effectively prevent diseases

 B. the inability to develop a successful vaccine

 C. owing to limited success in penetrating the household market

4. How should the biotechnology sector foster its competitiveness?

 A. by strengthening its research capabilities

 B. by making the industry a potential engine of sustained economic growth

 C. by raising awareness of the need to effectively prevent diseases

5. Where are examination reagents, medical supplies and medicaments key areas for development?

 A. in the biotech sector

 B. in the pharmaceutical industry

 C. in the community

G Reading Comprehension I

Pick the work or expression whose meaning is closest to the meaning of the underlined word or expression in the following passages.

Situation 1

1. E-learning in Taiwan has recently emerged as a highly promising learning <u>medium</u> for enhancing traditional classroom instruction.

   A. obstacle

   B. barrier

   C. approach

2. For example, e-learning <u>eliminates</u> time constraints and space limitations faced by classroom instruction.

   A. eradicates

   B. complements

   C. supplements

3. For example, e-learning eliminates time <u>constraints</u> and space limitations faced by classroom instruction.

   A. restraints

   B. openings

   C. opportunities

4. Another <u>strength</u> of e-learning is its ability to construct personal learning environments.

   A. detriment

   B. handicap

   C. merit

5. Internet-based English learning websites are a notable <u>example</u>.

   A. specimen

   B. anomaly

   C. deviance

6. Such websites commonly adopt the sharable content object reference model (SCORM), which emphasizes reusability, <u>accessibility</u>, durability and interoperability.

   A. impingement

   B. obtrusion

   C. openness

7. Such websites commonly adopt the sharable content object reference model (SCORM), which emphasizes reusability, accessibility, <u>durability</u> and interoperability.

   A. fragility

   B. firmness

   C. flimsiness

8. Successful adoption of the SCORM standard has significantly reduced the <u>temporal</u> and spatial constraints faced by e-learners.

   A. fleeting

   B. perennial

   C. enduring

9. Unsurprisingly, recent statistics indicate strong <u>growth</u> in e-learning.

   A. degradation

   B. deterioration

   C. evolution

10. For instance, in 2003, the e-learning market in the United States <u>generated</u> revenues of US$ 400,000,000, and the compound annual growth rate for e-

learning revenues is predicted to reach 20.7% from 2002 to 2007.

A. expended

B. produced

C. exhausted

11. In sum, e-learning represents a highly promising area for digital content <u>applications</u>.

A. excavations

B. implementations

C. extractions

12. Commercial <u>strategies</u> involving Internet-based English learning websites thus are receiving increased attention.

A. evasions

B. elusions

C. blueprints

13. Successful commercialization requires prioritizing market <u>orientation</u>, making teaching design and pedagogical content a primary concern.

A. egression

B. direction

C. migration

14. Interactive websites <u>enable</u> learners to enhance their competitiveness in school and at work.

A. empower

B. denigrate

C. tarnish

15. Successful learning websites adopt the latest information technologies while integrating the <u>expertise</u> of multidisciplinary professionals.

A. impotence

B. decrepitude

C. adeptness

16. For instance, several Internet-based English learning websites <u>employ</u> professionals in art design, information technology, marketing and curriculum design.

A. lay off

B. hire

C. suspend

17. Since potential revenues increase with market size, Internet-based English learning websites should <u>compete</u> not only for the Taiwan market, but also for the Greater China market.

A. abdicate

B. contend

C. repudiate

18. E-learning approaches and <u>related</u> expertise can also be applied for developing asynchronous learning and vocational training websites.

A. relevant

B. contrary

C. deviant

Situation 2

1. Since the establishment of the National Health Insurance scheme in 1995 and the National Health Information Network (HIN), medical-oriented information technology has been extensively adopted in clinical practice, as evidenced by <u>widespread</u> information system outsourcing.

A. insular

B. pervasive

C. parochial

2. An increasing number of hospitals and medical institutions purchase commercially produced medical information systems or <u>related</u> components.

A. dissimilar

B. allied

C. irrelevant

3. There is increasing <u>reliance</u> on information technology firms to design medical information systems and software.

A. assuredness

B. refutable

C. contestable

4. However, creating medical software involves several complex issues such as effectively <u>integrating</u> information.

A. dissipating

B. effusing

C. consolidating

5. While information technology firms can develop software to match individual hospital needs, integrating future applications into existing systems can be <u>problematic</u>.

A. dubious

B. contented

C. jocund

6. <u>Computerization</u> of hospital operations increased from 28% in 1994 to 57% in 1996.

A. handiwork

B. digitalization

C. workmanship

7. Available <u>medical</u> information systems include medical management systems, medical Intranet systems, Internet-based medical systems and electronic charts.

A. curative

B. officious

C. presumptuous

8. According to *Information Security Technical Report* (Vol.1, No3), the Outsourcing Institute forecasts annual <u>expenditures</u> on outsourcing by organizations in the United States at $100 billion dollars.

A. resources

B. nest egg

C. disbursements

9. Information technology (IT) outsourcing <u>accounted for</u> 40% of this total, or $40 billion dollars.

A. amounted to

B. divided into

C. extracted

10. The US information <u>industry</u> outsourced over $US 1,755,000,000 to the medical services sector in 2001, increasing to $US 2,809,000,000 in 2005.

A. fabrication

B. expatriation

C. excision

11. As evidenced by an average annual compound growth rate of 12.5%, information systems outsourcing in the medical services sector in the United States is clearly <u>growing</u>.

A. waning

B. abating

C. mushrooming

12. The largest area for outsourcing <u>continues</u> to be information technology.

    A. desists

    B. resumes

    C. refrains

13. The items most likely to be outsourced are hardware <u>maintenance</u>, training, applications development, re-engineering and mainframe data centers.

    A. oversight

    B. upkeep

    C. disregard

14. Therefore, the main areas in terms of marketing and direct sales to companies are servers, software applications, maintenance, networks, desktop systems and <u>end-user</u> support items.

    A. consumer

    B. wholesaler

    C. distributor

Situation 3

1. The biotech sector has emerged as one of the main areas of industrial growth in the new century owing to its <u>profound</u> impact on human quality of life.

    A. paltry

    B. negligible

    C. significant

2. As part of its efforts to prioritize this area in national development strategies, the Taiwanese government has offered numerous incentives for investment in and development of this industry, making the industry a <u>potential</u> engine of sustained economic growth.

    A. prospective

B. unpropitious

C. inauspicious

3. As part of its efforts to prioritize this area in national development strategies, the Taiwanese government has offered numerous <u>incentives</u> for investment in and development of this industry, making the industry a potential engine of sustained economic growth.

A. impetuses

B. impediment

C. determent

4. In the biotech sector, examination reagents, medical supplies and <u>medicaments</u> are all key areas for development.

A. inflictions

B. maladies

C. remedies

5. Additionally, the SARS crisis in 2003 <u>heightened</u> concerns over Taiwan's inability to effectively prevent epidemics and treat infected patients.

A. mitigated

B. augmented

C. abated

6. Efforts in drug and vaccine development following the SARS <u>epidemic</u> demonstrated the increased attention being paid to biotechnology.

A. pandemic

B. affluence

C. opulence

7. Besides raising awareness of the need to effectively prevent diseases or develop a successful vaccine, the SARS crisis <u>emphasized</u> the need to promote immunity in the community.

A. demoralized

B. played down

C. punctuated

8. While biotechnology efforts in Taiwan are focused on the pharmaceutical industry, the local market scale is small owing to limited success in penetrating the household market, as evidenced by the lack of research and development capabilities, <u>incomplete</u> clinical research and product trials, as well as the daunting regulations governing the use of specific drugs.

A. plenary

B. fragmentary

C. integral

9. While biotechnology efforts in Taiwan are focused on the pharmaceutical industry, the local market scale is small owing to limited success in penetrating the household market, as evidenced by the lack of research and development capabilities, incomplete clinical research and product trials, as well as the <u>daunting</u> regulations governing the use of specific drugs.

A. audacious

B. insipid

C. bland

10. To foster its competitiveness, the biotechnology sector should strengthen its research capabilities, develop patented technologies and attract biotechnology professionals with <u>expertise</u> in multidisciplinary fields.

A. incompetence

B. callowness

C. adeptness

H Common elements in forecasting market trends 預測市場趨勢 include the following elements:

1. Briefly state the rationale for developing this product.

   開發此產品的原因，科技層面：(1)相關產品；(2)潛在利潤

   · As an innovative technology, solar energy power is extensively developed in many industrialized countries. In this area, efficient production of solar energy panels is essential to generate electricity from the sun.

   · Living standards in Taiwan have dramatically increased in recent years, as evidenced by average per capita incomes surpassing $US 10,000 and the island's position among five of the largest holders of foreign exchange reserves worldwide.

   · A growing elderly population will also lead to an increase in the rate that individuals retire, with a shrinking younger generation to support the elderly. Therefore, understanding how the quality of life of the elderly and the retail housing market for this age group are related has received increasing attention given current trends in Taiwan's aging society.

2. Point out the market trends in the area of development while describing the financial aspect of developing this product

   財務層面：開發領域的市場趨勢

   · According to 2004 market statistics, anticancer drugs were among twenty one of the top selling 200 medicines, generating revenues of $US15 billion dollars. Accounting for 5% of medical market revenues, anticancer drugs will generate $US35.5 billion dollars in profit by 2010 at an annual growth rate of 7%.

· Given the increasing number of elderly and diabetic individuals, the market demand for curative products that can treat difficult-to-heal wounds is growing, as indicated by forecasted global market revenues to range from $US4,200,000,000 to 6,400,000,000 dollars by 2009.

· As the market and technology for biotechnology-based cosmetics continuously develop, the Industrial Technology Research Institute indicated that revenues from the local cosmetics sector ranged from 1,700,000,000 to 1,800,000,000 U.S. dollars in 2002.

3. Visually depict the organization's strategy and approach to developing the product.

產品開發的組織策略和方案：(1)策略的界定（包括最高目標及主要重點）；(2)執行策略方案的界定

· Effectively promoting the use of solar energy requires an emphasis not only on its continuous development to gain public confidence in its ability to provide for energy consumption in the future, but also on its applicability in a diverse array of product appliances and facilities.

· After manufacturing facilities are established in China and Japan, twenty chain stores and two branches will be established in those countries. In the next two years, $US 8,000,000 will be invested to establish three chain stores, along with an administrative department to determine the optimal locations for those stores.

· To foster its competitiveness, the domestic electronic communications sector must strengthen its research capabilities, develop lower electromagnetic wave technologies and attract technology professionals with expertise in multidisciplinary fields.

4. Summarize market survey results, particularly in terms of the product's commercial potential and categories of product application.

市場調查：(1)產品的商業潛力；(2)產品運用的範疇

· According to a recent market survey, although only around 20% of all disabled elderly in Taiwan receive institutional-based care, the market demand for institutional-based care is 30% (Department of Health, 1997). The 10% difference is equivalent to a market scale of at least 18,000 individuals. Moreover, the annual growth rate for disabled elderly in Taiwan is nearly 20 %.

· According to Taiwan's Annual Real Estate Report, an increasing number of individuals will live independently in rented apartment units located in housing complexes where the elderly live exclusively. While such a trend is aimed at upper income elderly, marketing efforts should be made to promote this future living trend among this growth sector in Taiwan and gain societal acceptance in a traditional society where the younger generation is normally expected to care for the elderly.

· With the number of knee osteoarthritis patients expected to increase from 15 million in 2000 to 19 million in 2010, i.e., an increase of 26%, viscosupplements are expected to undergo a surge in demand. Besides the enormous market potential for osteoarthritis medication, the market for post-surgical anti-adhesion barriers is also highly promising.

In the space below, forecast the market trends for a product or service.

_____

_____

_____

_____

_____

_____

_____

_____

_____

_____

Look at the following examples of forecasting marketing trends.

As an innovative technology, solar energy power is extensively developed in many industrialized countries. In this area, efficient production of solar energy panels is essential to generate electricity from the sun. As evidence of its commercial potential, solar energy panels are increasingly found in batteries and automobiles. The relative inexpensiveness of solar panels and their environmentally friendly use reflect the current trend to use renewable resources instead of non-renewable ones such as petroleum, especially given its growing retail price and scarcity. Solar panels are a particularly attractive alternative to petroleum in transportation. Competitions involving solar energy driven automobiles are common, subsequently raising public awareness of the latest advances and the advantages of solar energy. Encouraging an environmental consciousness among the general public is another effective means of raising the awareness of solar energy and its environmental benefits. However, in practice, solar energy panels still have certain limitations, such as use for extended periods or use in poor weather conditions, explaining the hesitancy among the general public to accept it as a viable alternative to non-renewable energy sources. Therefore, effectively promoting the use of solar energy requires an emphasis not only on its continuous development to gain public confidence in its ability to provide for energy consumption in the future, but also on its applicability in a diverse array of product appliances and facilities.

While offering a healthy snack alternative for consumers, homemade-style cookies can often be found in convenience stores and coffee shops. Made of wheat flour, eggs and sugar, homemade-style cookies also have other ingredients, including cocoa powder, coffee and whip cream - all of which are fresh and do not have chemical additives. Uni-President Corporation began collaborating with homemade-style cookie makers in 1999, providing those makers with a strong brand recognition. Additionally, a Taiwanese representative agent will promote and distribute these products in Asia. Initially, three chain stores will be established in Hsinchu, Kaohsiung and Taipei, with branch stores soon to follow island wide. After manufacturing facilities are established in China and Japan, twenty chain stores and two branches will be established in those countries. In the next two years, $US 8,000,000 will be invested to establish three chain stores, along with an administrative department to determine the optimal locations for those stores. Additionally, an operating budget of $US 1,000,000 will be allocated. Before 2010, generated revenues will hopefully reach $8,000,000, with a branch company to be established in China by 2012. Second, all chain stores will be established proportionately because market differentiation will enable homemade-style cookies to avoid price competition with other food products. Products image will be represented in its higher quality and retail price than other cookies have. Third, all products will adopt the advertising slogan, "The best healthy snacks insist on using the freshest ingredients" and "Insist on the best snacks." Commercial success is widely anticipated given the growing health consciousness in Asia, hopefully generating large profits in the near future.

According to 2004 market statistics, anticancer drugs ranked 21st among the top selling 200 medicines, generating revenues of $US15 billion dollars. Accounting for 5% of medical market revenues, anticancer drugs will generate $US35.5 billion dollars in revenues by 2010 at an annual growth rate of 7%. The World Health Organization forecasts an annual 1,086,000 reported cases of cancer, with 672,000 mortalities annually. Above statistics reflect the serious health and societal implications that cancer holds. Taiwan is no exception, with cancer ranking as the leading cause of death in Taiwan in 1993, or 36,357 fatalities. Among cancer types, lung cancer ranks first, followed by liver cancer. Taiwanese pharmaceuticals must thus pay close attention to the strong market potential of anticancer drugs. Taiwan also has a uniquely high incidence of liver cancer, which should be cause for concern. With nearly 200 cancer types worldwide, anticancer medicines can control the rate at which the disease spreads through daily medication. Among the commonly administered anticancer drugs include cytotoxic drugs, hormonal therapy, targeted therapy and genetic therapy. Additionally, Herceptin, Avastin, Rituxan, Tarceva, MabThera and Xeloda are among the latest prescribed anticancer medicines. Global pharmaceutical giants such as Roche, Genentech, Bayer, Pfizer, Novartis and AstraZeneca have expended considerable effort in developing such anticancer medicines. A notable trend is the market for VEGF inhibitors, which increased its growth rate by 10 times in recent years. Therefore, local pharmaceutical manufacturers must enhance their research capabilities to enter the lucrative anticancer drug market.

Situation 4

Situation 5

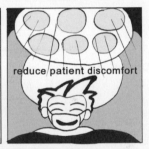

Situation 6

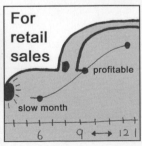
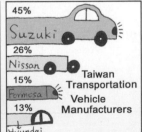

I Write down the key points of the situations on the preceding page, while the instructor reads aloud the script from the Answer Key. Alternatively, students can listen to the script online at www.chineseowl.idv. tw

Situation 4

_____

_____

_____

Situation 5

_____

_____

_____

Situation 6

_____

_____

_____

J Oral practice II

Based on the three situations in this unit, write three questions beginning with *How*, and answer them. The questions do not need to come directly from these situations.

Examples

How is EASYCARD unlike the conventional magnetic card?

It is the first IC card for use in mass transportation in Taiwan.

How are EASYCARD transactions executed wirelessly?

Through memory IC chips and induction circuits implanted in the card

1. _____

   _____

2. _____

   _____

3. _____

   _____

K Based on the three situations in this unit, write three questions beginning with **What**, and answer them. The questions do not need to come directly from these situations.

Examples

What has expanded owing to the increasing incidence of cancer?

The market demand for chemotherapy medicine

What has led to the popularization of preventive measures?

Public fear of cancer

1. _____

   _____

2. _____

   _____

3. _____

   _____

L Based on the three situations in this unit, write three questions beginning with **Why**, and answer them. The questions do not need to come directly from these situations.

Examples

Why is the installation of digitalized consumer products in automobiles in Taiwan popular?

Because of continuous technological advances

Why is future automobile design anticipated to integrate media entertainment with the vehicle computer system?

Because automobiles are gradually becoming communications centers

1. _____

   _____

2. _____

   _____

3. _____

   _____

M Write questions that match the answers provided.

_____

_____

Continuous technological advances

_____

_____

Because of their emphasis on catering to individual preferences and their recreational appeal

_____

_____

In 1993

N Listening Comprehension II

Situation 4

1. When was the Taipei Smart Card Corporation established?

  A. in 2001

  B. in 1999

  C. in 2000

2. How many private bus companies cooperated in the establishment of the Taipei Smart Card Corporation?

  A. 13

  B. 15

  C. 23

3. How are EASYCARD transactions executed?

  A. by increasing capacity, durability, speed, accuracy and security

  B. by enabling prompt transactions

  C. wirelessly through memory IC chips and induction circuits implanted in the card

4. How does EASYCARD differ from the conventional magnetic card?

  A. It is the first IC card for use in mass transportation in Taiwan

  B. It frees passengers from carrying coins or tokens, or making repetitive ticket purchases.?

  C. It conveniently combines payment for several transport modes into one ticket.

5. What does EASYCARD represent?

  A. a milestone for intelligent transport systems in Taiwan

  B. prompt transactions

  C. reduced traffic congestion

Situation 5

1. Why has the market demand for chemotherapy medicine expanded?

    A. owing to the need to avoid excessive drinking and smoking

    B. owing to the need to popularize preventive measures

    C. owing to the increasing incidence of cancer

2. What has led to the popularization of preventive measures?

    A. research efforts to develop more effective treatments

    B. public fear of cancer

    C. the need for a nutritionally balanced diet without greasy food

3. What are several biotech firms developing?

    A. foods with anti-cancer properties

    B. products to avoid excessive drinking and smoking

    C. exercise plans

4. Why have research efforts in chemotherapy medicine been accelerated?

    A. to encourage individuals to eat a nutritionally balanced diet and avoid greasy food

    B. to develop more effective treatments

    C. to avoid excessive drinking and smoking

5. What normally occurs after physical treatment and aims to remove all remaining cancer cells?

    A. chemotherapy

    B. organ removal

    C. radiotherapy

Situation 6

1. Where was the MARCH car brand launched in 1993?

    A. in the United States

B. in Japan

C. in Taiwan

2. When does the average car owner purchase a new model?

A. every five years

B. every ten years

C. every eight years

3. How have continuous technological advances popularized the installation of digitalized consumer products in automobiles in Taiwan?

A. through a Japanese parent factory

B. from telematics to rear seat entertainment systems

C. through ethnic appeal and numerous exterior accessories

4. What car brand is intended to appeal to younger car drivers?

A. March

B. Matiz

C. Solio

5. Which automobile corporation focuses on cars with less than 1000CC horsepower?

A. Formosa

B. Suzuki

C. Nissan

6. Which is the leading car brand in Taiwan in terms of production and sales?

A. Nissan

B. Hyundai

C. Suzuki

O Reading Comprehension II

Pick the work or expression whose meaning is closest to the meaning of the underlined word or expression in the following passages.

Situation 4

1. Established in 2000, the Taipei Smart Card Corporation (TSCC) was <u>set up</u> by cooperation among the Taipei City Government, the Taipei Rapid Transit Corporation, 13 private bus companies, TAIPEIBANK, Mitac Inc., Taishin International Bank, and various others.

   A. disintegrated

   B. contrived

   C. dissolved

2. TSCC has <u>implemented</u> a contact free smart-card ticketing system for buses, the metro and public off-road car parks in Taipei.

   A. promulgated

   B. desist

   C. suspend

3. The establishment of TSCC marks a <u>milestone</u> for intelligent transport systems in Taiwan.?

   A. juncture

   B. tediousness

   C. ho-hum

4. In the future, TSCC will <u>expand</u> its services to other areas. Unlike the conventional magnetic card, EASYCARD is the first IC card for use in mass transportation in Taiwan.

   A. depreciate

B. deplete

C. aggrandize

5. EASYCARD transactions are executed wirelessly through memory IC chips and induction circuits <u>implanted</u> in the card.?

A. extracted

B. embedded

C. plucked out

6. With features of large capacity, <u>durability</u>, speed, accuracy and security, EASYCARD enables prompt transactions and has a long lifetime.

A. instability

B. immutability

C. fluctuation

7. EASYCARD frees passengers from carrying coins or tokens, or making <u>repetitive</u> ticket purchases.?

A. recurrent

B. unswerving

C. undeviating

8. Additionally, transfer rides do not require advance ticket validation, thus enhancing user <u>convenience</u>.

A. inopportune

B. expedience

C. cumbersomeness

9. EASYCARD <u>conveniently</u> combines payment for several transport modes into one ticket.

A. rigidly

B. flexibly

C. obstinately

10. Finally, EASYCARD represents reduced traffic <u>congestion</u>, as public transport utilization rates increase in response to improved service quality.?

A. dispersion

B. dissipation

C. crowding

Situation 5

1. Market demand for chemotherapy medicine has expanded owing to the increasing <u>incidence</u> of cancer, accelerating research efforts to develop more effective treatments.

A. aversion

B. prevention

C. occurrence

2. Besides chemotherapy for treating and <u>preventing</u> cancer, many food products with anti-cancer claims have recently been marketed.

A. granting

B. inhibiting

C. allocating

3. Public fear of cancer has led to the popularization of preventive <u>measures</u>.

A. paradigms

B. reluctance

C. repugnance

4. For example, individuals who wish to prevent cancer are encouraged to avoid <u>excessive</u> drinking and smoking, get sufficient sleep, exercise, eat a nutritionally balanced diet and avoid greasy food.

A. temperate

B. moderate

C. exorbitant

5. Additionally, several biotech firms are developing foods with anti-cancer properties.

   A. characteristics

   B. aberration

   C. anomaly

6. Successful product commercialization will undoubtedly yield numerous benefits.

   A. categorically

   B. tentatively

   C. speculatively

7. Cancer treatments include a) physical treatment, through organ removal and subsequent radiotherapy b) chemotherapy, which normally occurs after physical treatment and aims to remove all remaining cancer cells.

   A. annexing

   B. appending

   C. extrication

8. Cancer treatments include a) physical treatment, through organ removal and subsequent radiotherapy b) chemotherapy, which normally occurs after physical treatment and aims to remove all remaining cancer cells.

   A. lingering

   B. incipient

   C. commencing

9. Regarding trends in the development of chemotherapy medicine, more effective chemotherapy drugs developed in the future may be able to prevent metastasis in cancer cells.

   A. inclinations

   B. hesitations

C. evasion

10. Such drugs would greatly reduce patient <u>discomfort</u> and the adverse effects of physical treatment during cancer therapy.

    A. tranquility

    B. solace

    C. malaise

Situation 6

1. Continuous technological advances have popularized the <u>installation</u> of digitalized consumer products in automobiles in Taiwan: from telematics to rear seat entertainment systems.

    A. removal

    B. placement

    C. eradication

2. Automobiles are gradually becoming communications centers, with future automobile design anticipated to <u>integrate</u> media entertainment with the vehicle computer system.

    A. disentangle

    B. consolidate

    C. uncouple

3. Rear seat multimedia entertainment systems are a highly <u>promising</u> area for further technological development and are attracting increased commercial interest.

    A. prospective

    B. dismal

    C. morose

4. The MARCH car brand was <u>launched</u> in Taiwan in 1993, largely owing to the technological limitations of its Japanese parent factory in manufacturing mini-sized cars.

　　A. recanted

　　B. retracted

　　C. impelled

5. Another local car brand, Matiz, from the Formosa Automobile Corporation, focuses on cars with less than 1000CC horsepower, with a <u>diverse</u> array of color combinations.

　　A. homogeneous

　　B. heterogeneous

　　C. analogous

6. Meanwhile, the Solio car brand from Suzuki is intended to <u>appeal</u> to younger car drivers with its ethnic appeal and numerous exterior accessories, such as fins.

　　A. implore

　　B. intimidate

　　C. appall

7. As for retail sales, June is generally a slow month, with the most <u>profitable</u> period running from September through December or January.

　　A. restricted

　　B. strained

　　C. lucrative

8. The average car owner <u>purchases</u> a new model every ten years.

　　A. liquidates

　　B. remits

　　C. procures

9. According to January 2005 statistics on automobile sales from the Taiwan Transportation Vehicle Manufacturers Association, the <u>leading</u> car brands in Taiwan in terms of production and sales are Suzuki (45%), Nissan (26%), Formosa (15%) and Hyundai (13%).

A. paramount

B. flagging

C. decelerating

10. Japanese brands thus are highly successful in Taiwan, with their emphasis on catering to individual <u>preferences</u> and their recreational appeal.

A. negation

B. partiality

C. renunciation

11. Restated, an appealing car exterior and large variety of interior accessories with driving and entertainment functions are the <u>foundations</u> of the commercial success of these Japanese car brands.

A. fundamentals

B. assumptions

C. hypotheses

Unit Two

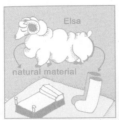

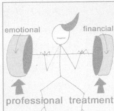

Describing Product or Service Development

產品或服務研發

1. Introduce the current status of product or service development.
   產品開發的現階段狀況

2. Describe its market value.
   市場的評價

3. Point out unique features and characteristics of product or service development.
   產品開發的獨創性和特色

4. List major manufacturers of this product or service in Taiwan.
   台灣同類產品的主要製造廠商

5. Explain the rationale for further product or service development in this area.
   在既有領域下未來產品開發的原因

Vocabulary and related expressions

consumer evaluations	消費者評估
market niche	市場利基
turbulence of daily life	每日生活的騷亂
exceed consumer expectations	超過消費者預期
highly competitive market	高度競爭的市場
long term care	長期的照護
abolished	廢除
discrepancy	不一致之處
alleviate	減輕
wound severity	傷口的嚴重性
increasing life expectancies	日漸增加的預期壽命
eradication	根除
chronic illnesses	慢性病
therapeutic treatment	治療
alleviates the burden	減輕負擔
National Health Insurance resources	國家健保資源
immensely popular	廣大地受歡迎
erode	腐蝕
dominant market share	占優勢的市場占有率
technology cooperation	科技合作
improved living standards	增加生活水準
routine physical examination	慣例的身體檢查
precautions	預防／警惕
cure rates	治癒率
optimizing	使完美
prescribed dosage	處方（藥的）劑量

Situation 1

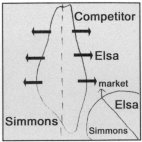

Situation 2

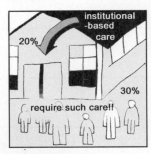
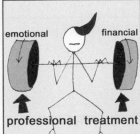

Situation 3

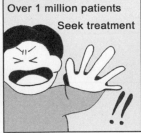
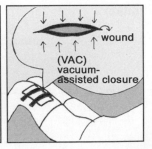

A Write down the key points of the situations on the preceding page, while the instructor reads aloud the script from the Answer Key. Alternatively, students can listen to the script online at www.chineseowl.idv. tw

Situation 1

_____

_____

Situation 2

_____

_____

_____

Situation 3

_____

_____

_____

B Oral practice I

Based on the three situations in this unit, write three questions beginning with *When*, and answer them. The questions do not need to come directly from these situations.

Examples

When did Elsa begin striving to satisfy customers?

Decades ago

When did the German mattress brand Elsa enter the Taiwan market?

In 2004

1. _____

   _____

2. _____

   _____

3. _____

   _____

C Based on the three situations in this unit, write three questions beginning with **What**, and answer them. The questions do not need to come directly from these situations.

Examples

What is the annual growth rate of disabled elderly in Taiwan?

Nearly 20%

What emerged in Taiwan in the late 1980s?

Long term care

1. _____

   _____

2. _____

   _____

3. _____

   _____

D Based on the three situations in this unit, write three questions beginning with *How*, and answer them. The questions do not need to come directly from these situations.

Examples

How is wound care treatment crucial in nursing care?

It involves the assessment of wound severity and appropriate treatment.

How do medical personnel benefit not only patients, but also their relatives and society as a whole?

By striving to heal patients with wounds by adopting the latest therapeutic treatment strategies

1. _____

   _____

2. _____

   _____

3. _____

   _____

E Write questions that match the answers provided.

1. _____

_____

In the United States

2. _____

_____

It reduces fatality rates, inhibits disease growth during the early stages, alleviates the burden on relatives in terms of manpower and financial resources and reduces hospital expenditures, ultimately reducing already strained National Health Insurance resources.

3. _____

_____

Over the past three decades

F Listening Comprehension I

Situation 1

1. What makes Taiwan an ideal market for the diverse products Elsa manufactures?

   A. Its cold and damp winters

   B. high consumer expectations

   C. Its elevated standard of living

2. Who does Simmons offer quality mattresses for?

   A. consumers with a humidifier in their bedrooms

   B. individuals with high consumer expectations

   C. individuals who have trouble sleeping

3. When does Elsa plan to team up with Teco Company?

   A. in 2005

   B. in 2006

   C. in 2007

4. What does Simmons educate consumers to do?

   A. deal with the turbulence of daily life

   B. combine innovation with comfort

   C. maximize their sleeping experience and ensure a healthy lifestyle

5. Why should consumers sleep more comfortably and healthily given the damp climate in Taiwan?

   A. They have a stronger purchasing power than previously.

   B. They have a humidifier in their bedrooms.

   C. They know how to deal with the turbulence of daily life.

Situation 2

1. When did long term care emerge in Taiwan?

   A. in the late 1990s

   B. in the late 1980s

   C. in the early 1980s

2. What ushered in the rapid growth of institutional-based organizations?

   A. strong public pressure

   B. intense market competition

   C. the Senior Citizens Welfare Law

3. How many disabled elderly individuals represent the market demand for long term institutional-based care in Taiwan?

   A. more than 18,000

   B. at least 18,000

   C. nearly 18,000

4. Which group accounts for 69% of all care providers for the disabled elderly in Taiwan?

   A. family members without professional training

   B. family members with professional training

   C. manpower organizations

5. What is the annual growth rate of disabled elderly in Taiwan?

   A. nearly 20%

   B. nearly 25%

   C. nearly 30%

Situation 3

1. Which group is increasingly emphasizing the need to reduce wound treatment associated costs in clinical practice?

    A. Taiwanese physicians in clinical practice

    B. Taiwan's National Health Insurance Bureau

    C. Taiwanese hospital administrators

2. How many patients in the United States alone seek treatment annually for chronic wounds?

    A. nearly 1 million

    B. roughly 1 million

    C. over 1 million

3. What is vacuum-assisted closure (VAC)?

    A. a recently developed clinical treatment procedure for patients with chronic illnesses

    B. a recently developed wound management procedure

    C. a recently developed hospital expenditures procedure

4. What can ultimately reduce already strained National Health Insurance resources?

    A. decreasing monthly premiums

    B. lowering medical costs

    C. reducing wound healing times

5. What does VAC apply to a wound?

    A. constant pressure

    B. negative pressure

    C. a porous, open-cell foam

G Reading Comprehension I
Pick the work or expression whose meaning is closest to the meaning of the underlined word or expression in the following passages.

Situation 1

1. With its strong emphasis on using natural materials, the German mattress brand Elsa has ranked highly in consumer <u>evaluations</u> since entering the Taiwan market in 2004.

   A. dishevel

   B. disarray

   C. appraisals

2. Established in Germany in 1924, Elsa has <u>strived</u> to satisfy customers for decades.

   A. renounced

   B. relinquished

   C. aspired

3. The cold and damp winters in Taiwan make the island an <u>ideal</u> market for the diverse products Elsa manufactures, including woolen carpets, blankets, socks, stockings and nightclothes.

   A. exemplary

   B. flawed

   C. deficient

4. The main competitor of Elsa in the Taiwan market is Simmons Mattress Company of the United States, which began mass producing spring beds in 1876 and built up a <u>firm</u> position in the Taiwan market.

   A. faltering

   B. impermeable

C. precarious

5. Simmons is the leader in its market niche, and focuses on eliminating the stress of shopping for mattresses by combining <u>innovation</u> with comfort.

A. reversion

B. metamorphosis

C. relapse

6. Simmons seeks to assure customers that they are purchasing a quality mattress, and also to <u>convince</u> them quality sleep is essential for dealing with the turbulence of daily life.

A. persuade

B. sidetrack

C. deter

7. For individuals who have <u>trouble</u> sleeping, Simmons offers quality mattresses that often exceed consumer expectations.

A. contentment

B. tranquility

C. aggravation

8. Elsa faces a challenge in competing with a <u>well entrenched</u> rival like Simmons.

A. well fortified

B. well concealed

C. well covered

9. Knowledge is a <u>key</u> concern.

A. minor

B. primary

C. slight

10. Rather than merely selling quality mattresses, Simmons educates <u>consumers</u> to maximize their sleeping experience and ensure a healthy lifestyle.

A. merchants

B. purchasers

C. wholesalers

11. In 2005, Elsa plans to <u>team up</u> with Teco Company and offer consumers a 50% discount off of an Elsa mattress when purchasing a Teco humidifier.

A. collaborate

B. compete

C. rival

12. Given the <u>damp</u> climate in Taiwan, consumers with a humidifier in their bedrooms should sleep more comfortably and healthily.

A. soggy

B. desiccated

C. parched

13. Such a <u>partnership</u> should help improve Elsa's position in this highly competitive market.

A. contention

B. rivalry

C. collusion

Situation 2

1. Long term care <u>emerged</u> in Taiwan in the late 1980s.

A. recoiled

B. emanated

C. evacuated

2. A turning point occurred in 1997 with the passage of the Senior Citizens Welfare Law, which placed unregistered healthcare institutes under <u>pressure</u> and eventually saw them abolished by 2000.

A. tension

B. repose

C. tranquility

3. This legislation ushered in the rapid growth of institutional-based organizations from 1998, with <u>stable</u> growth expected well beyond 2000.

A. anchored

B. precarious

C. tottery

4. Competition among institutional-based long term care facilities currently is <u>fierce</u>.

A. happy-go-lucky

B. jocund

C. relentless

5. A <u>recent</u> market survey indicated that while only around 20% of all disabled elderly in Taiwan receive institutional-based care, 30% of the disabled elderly in Taiwan require such care (Department of Health, 1997).

A. archaic

B. antiquated

C. newfangled

6. This <u>discrepancy</u> represents a market demand of at least 18,000 individuals.

A. variance

B. uniformity

C. symmetry

7. To meet this <u>demand</u>, Taiwan has relied on small-scale institutional care facilities.

A. alternative

B. mandate

C. equivalent

8. Such facilities have become <u>popular</u> for four reasons. First, family members without professional training account for 69% of all care providers for the disabled elderly in Taiwan.

A. rebuffed

B. in vogue

C. objectionable

9. This situation creates high emotional and financial stress, and providing professional treatment to the disabled elderly can greatly <u>alleviate</u> family tensions.

A. abate

B. aggrandize

C. amplify

10. Second, the annual growth <u>rate</u> of disabled elderly in Taiwan is nearly 20%.

A. reversal

B. regression

C. pace

11. Third, modern lifestyles and urbanization have significantly <u>transformed</u> familial patterns, as reflected by the tendency of adults to live apart from their parents and offer their parents less assistance than previously.

A. metamorphosed

B. lingered

C. dwelled

12. Meanwhile, increasing daily pressures in daily life, family interactions, and the growing female workforce have reduced numbers of non-professional <u>caretakers</u> for the disabled elderly.

A. stewards

B. detractors

C. distracters

13. Consequently, <u>professional</u> care givers are increasingly important for meeting demand.

    A. amateurish

    B. incompetent

    C. polished

14. Fourth, a clear discrepancy in supply of institutional-based long term care facilities exists between <u>urban</u> and rural areas.

    A. rustic

    B. metropolitan

    C. agrarian

15. While demand for such facilities is lower in rural areas, marketing <u>opportunities</u> still exist for smaller scale facilities.

    A. barricades

    B. contingencies

    C. impediments

Situation 3

1. Wound care treatment is crucial in nursing care, and involves the assessment of wound severity and <u>appropriate</u> treatment.

    A. out of character

    B. unsuited

    C. adequate

2. Increasing life expectancies globally over the past three decades, a growing elderly population and the <u>eradication</u> or alleviation of many systemic diseases have all contributed to the urgent need to clinically treat patients with chronic illnesses, especially those with difficult to heal wounds.

    A. tectonics

B. annihilation

C. fabrication

3. While striving to heal patients with wounds by adopting the latest <u>therapeutic</u> treatment strategies, medical personnel benefit not only patients, but also their relatives and society as a whole.

A. cataclysmic

B. incendiary

C. curative

4. Reducing wound healing times reduces fatality rates, <u>inhibits</u> disease growth during the early stages, alleviates the burden on relatives in terms of manpower and financial resources and reduces hospital expenditures, ultimately reducing already strained National Health Insurance resources.

A. represses

B. instigates

C. provokes

5. Statistics <u>demonstrate</u> the severity of this problem.

A. camouflage

B. conceal

C. substantiate

6. In the United States alone, over 1 million patients seek treatment annually for chronic wounds, with treatment costs <u>totaling</u> several hundred million dollars.

A. subtracting

B. rescinding

C. accumulating

7. Expenses associated with length of hospital stay and the <u>extent</u> of wound care treatment are valuable indexes of the severity of the wound treatment problem.

A. fraction

B. magnitude

C. segment

8. Thus, Taiwanese hospital administrators are increasingly <u>emphasizing</u> the need to reduce wound treatment associated costs in clinical practice.

A. squelching

B. shunting

C. punctuating

9. A recently developed wound management procedure, vacuum-assisted closure (VAC), applies <u>negative</u> pressure to a wound through a porous, open-cell foam that fills the wound cavity.

A. nullifying

B. supplemental

C. additional

10. The <u>advantages</u> include rapid wound healing, reduced pain, shorter hospital stays, lower medical costs and less need for nursing care.

A. merits

B. liability

C. hindrance

11. This procedure can also be applied to patients with multiple wounds, as well as to <u>recurring</u> wounds suffered by many elderly patients.

A. preventive

B. incessant

C. deterrent

H Common elements in describing product or service development產品或服務研發include the following:

1. Introduce the current status of product or service development.
   產品開發的現階段狀況

   · Despite the increasing number of solar energy construction projects in Taiwan, island wide development of this energy alternative is still in its preliminary phase, making it still impractical for satisfying daily consumption needs. Still, overseas solar energy firms have successfully transferred relevant technologies that will ultimately make Taiwan less dependent on non-renewable energy sources.
   · In line with trends in the local pharmaceutical sector, large-scale food manufacturers heavily invest in a diverse product line of health foods to satisfy consumer demands of better taste quality and a competitive retail price.
   · First adopted in Japan in 1995 and then in Taiwan in 2001 by First International Telecom, PHS mobile phones address the safety concerns of handset users with its use of low electromagnetic wave, subsequently gaining a significant competitive edge in the local retail market.

2. Describe its market value.
   市場的評價

   · With generated revenues of 8 billion New Taiwanese dollars annually, the health food market in Taiwan is forecasted to expand nearly twice in the near future.
   · The number of cosmetic surgeries and treatments increased five folds over a one-year period (2003-2004): from approximately 30,000 to more

than 150,000.

· Market growth has shifted from industrialized countries to developing ones in South America, Eastern Europe and Asia (especially China), with the compound annual growth rate exceeding 10.4% over the past five years.

3. Point out unique features and characteristics of product or service development.

產品開發的獨創性和特色

· As nearly 70% of these products are imported from abroad, the market opportunities for domestic manufacturers are immense, with the added potential of developing food products that are more conducive to local tastes.

· While Taiwan's automotive industry is unable to replace imported models directly, locally producing a popular imported brand such as the Accord RP-WC is a viable alternative while, at the same time, satisfying consumer demand.

· The strong demand for cosmetic surgery closely corresponds not only to the public perception that an attractive appearance is vital for professional and social settings, but also to the beliefs of many Taiwanese that one's face or body can influence one's fate, as postulated by Chinese geomantic theory.

4. List major manufacturers of this product or service in Taiwan.

台灣同類產品的主要製造廠商

· With Suzuki, Nissan, Formosa and Hyundai Motor Corporations taking the initiative in Taiwan, automobile manufacturing has approached a certain degree of maturity given the advanced production technologies adopted, variety in exterior and the latest product functions.

· As the main competitor of PHS in the island's mobile phone market, Chunghwa Telecom Company, Taiwan Mobile Company and Far Eastone Telecommunications Company have gained a market niche through their sales promotional strategies that cater to customer needs in order to ensure flexible and reliable communication.

· Several collaborative efforts are underway among multidisciplinary experts to further develop Chinese herbal medicine, as evidenced by the growing number of technology transfers from research institutes, teaching hospitals and clinical testing centers to local industry for commercialization.

5. Explain the rationale for further product or service development in this area.

在既有領域下未來產品開發的原因

· Widespread commercialization of solar energy technology depends on the ability to make it convenient and accessible, while popularizing its appeal among environmentally conscious individuals.

· Continued growth of this sector largely depends on the ability of local manufacturers to continuously adopt new manufacturing technologies and offer variety in exterior features and digital products in the interior.

· Continuously developing this market sector depends on the ability to adopt the latest technical advances in production and food preparation, with particular emphasis on the following: adoption of the latest verification technologies such as toxicity testing to ensure food safety, analysis of the incorporation of unique ingredients such as those coming from Chinese herbal medicine, promotion of biotechnology-based health care drinks, utilization of molecular biotechnology and fermented technology to develop new food products and enhancement of extraction and purification procedures during food processing.

In the space below, describe the development of a product or a service.

_____

_____

_____

_____

_____

_____

_____

_____

_____

Look at the following examples of describing product or service development.

Despite the increasing number of solar energy construction projects in Taiwan, island wide development of this energy alternative is still in its preliminary phase, making it still impractical for satisfying daily consumption needs. Still, overseas solar energy firms have successfully transferred relevant technologies that will ultimately make Taiwan less dependent on non-renewable energy sources. Of the six Taiwanese manufacturers in this area, sufficient expertise and modernized facilities are still lacking in comparison with their overseas counterparts. Despite their lack of experience, these manufactories continuously make progress. Given the skyrocketing price of petroleum, solar energy has received considerable attention, as evidenced by the growing receptiveness among environmentally conscious individuals worldwide towards this viable alternative given its environmentally friendliness and relative inexpensiveness. Still, development of solar energy technologies is in its infant stage of development in Taiwan since few enterprises are devoted to this area. A notable exception is Motech Corporation, which is committed to developing the potential of solar energy technologies. Solar energy can be used in many electronic products, including computers, heaters and even cars. With its consumer appeal as an economic and clean substitute for petroleum products in producing electricity, solar energy will be increasingly adopted in 3C products, such as notebook computers and PDAs. In sum, widespread commercialization of solar energy technology depends on the ability to make it convenient and accessible, while popularizing its appeal among environmentally conscious individuals.

While accounting for 10% of the market share for homemade-style cookies in Taiwan, ABC Company is determined to increase its revenues by identifying potential customers that would normally not be considered. Besides core customers, consumers can be classified as first level non-customers, second level non-customers or third level non-customers. ABC Company attempts to devise a market niche for each of these groups, Nearest to the core customers, first level non-customers will soon stop purchasing ABC's products. Although a consumer of ABC's products, this group will purchase other products if they are more appealing since so many snacks are available for purchase. ABC Company should thus develop a more unique product line in terms of a broader array of tastes and flavors, thus offering customers with a wider range of choice. Successfully implementing this strategy could perhaps increase ABC's market share to 18%. How to develop a consumer market for second level non-customers is more difficult since they would normally refuse to purchase ABC products, largely owing to the high calorie content of cookies and fear of obesity. ABC Company should thus develop low-calorie cookies and, in doing so, further increase its market share to 27%. Finally, third level non-customers represent a market segment that has not yet been identified and tapped into. Given that their potential has never been developed, the demands of this group differ from those of others. ABC Company should thus develop other snack varieties, including breads, cakes and candy. Successful implementation of this marketing strategy would further increase ABC's market share to 40%. With the importance of effective marketing strategies, companies should pay more attention to the demands of non-customers.

Growing consumer demand for health food that is nutritious and tasty reflects the increasing health consciousness among Taiwanese residents. In line with trends in the local pharmaceutical sector, large-scale food manufacturers heavily invest in a diverse product line of health foods to satisfy consumer demands of better taste quality and a competitive retail price. Given the unique ingredients of health food, Taiwanese manufacturers adopt biotechnology to ensure its high quality. For instance, DNA chip and protein chip technologies are used to enrich the ingredients of health food. With generated revenues of $NT8 billion annually, the health food market in Taiwan is forecasted to expand nearly twice in the near future. As nearly 70% of these products are imported from abroad, the market opportunities for domestic manufacturers are immense, with the added potential of developing food products that are more palatable with local tastes. Advances in biotechnology should be further exploited to produce high quality health food that offers diversity and addresses hygienic concerns over storage and freshness. Continuously developing this market sector depends on the ability to adopt the latest technical advances in production and food preparation, with particular emphasis on the following: adoption of the latest verification technologies such as toxicity testing to ensure food safety, analysis of the incorporation of unique ingredients such as those coming from Chinese herbal medicine, promotion of biotechnology-based health care drinks, and utilization of molecular biotechnology and fermented technology to develop new food products and enhance extraction and purification procedures during food processing.

Situation 4

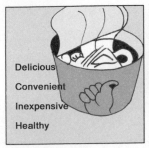

Situation 5

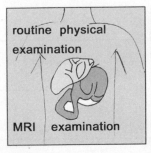

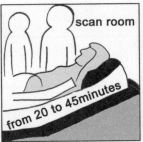

Situation 6

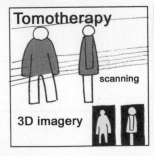

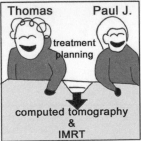

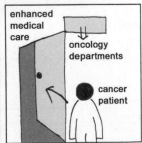

| I | Write down the key points of the situations on the preceding page, while the instructor reads aloud the script from the Answer Key. Alternatively, students can listen to the script online at www.chineseowl.idv. tw |

Situation 4

_____

_____

_____

Situation 5

_____

_____

_____

Situation 6

_____

_____

_____

J Oral practice II.
Based on the three situations in this unit, write three questions beginning with *Why*, and answer them. The questions do not need to come directly from these situations.

Examples

Why have instant noodles been a staple food item among Taiwanese for more than four decades?

They are delicious, convenient, inexpensive and healthy.

Why did local producers begin to gradually erode Japan's market dominance?

Because Wei Lih Food Manufacturers established a food processing plant in Changhua in 1970

1. _____

_____

2. _____

_____

3. _____

_____

K Based on the three situations in this unit, write three questions beginning with ***What***, and answer them. The questions do not need to come directly from these situations.

Examples

What has made Taiwanese more health conscious and recreation-oriented?

Improved living standards

What do most individuals pay extra for?

An MRI examination when undergoing their routine physical examination

1. _____

   _____

2. _____

   _____

3. _____

   _____

L Based on the three situations in this unit, write three questions beginning with *How*, and answer them. The questions do not need to come directly from these situations.

Examples

How can tomotherapy offer the most advanced radiation delivery system available?

Through its enhanced dose modulation and accurate targeting of specific locations

How can tomotherapy allow physicians to verify treatment volumes in advance?

Through 3D imagery via TomoImage scanning

1. _____

   _____

2. _____

   _____

3. _____

   _____

M Write questions that match the answers provided.

_____

_____

Tomotherapy

_____

_____

Professor Thomas Rockwell Mackie and the mathematician and software engineer Paul J. Reckwerdt at the University of Wisconsin-Madison ten years ago

_____

_____

Through using a unique verification CT to confirm the tumor position before each treatment, enabling precise delivery of the radiation dosage

N Listening Comprehension II

Situation 4

1. When did the production of instant noodles in Taiwan begin?

   A. in 1970

   B. in 1967

   C. in 1968

2. What company established a food processing plant in Changhua in 1970?

   A. Uni-President Enterprises

   B. Vedan Enterprise Corporation

   C. Wei Lih Food Manufacturers

3. Which age group is the biggest consumer of instant noodles?

   A. 7-14 year olds

   B. 30-45 year olds

   C. 15-29 year olds

4. How much in revenues did Taiwanese manufacturers of instant noodles achieve in 2002?

   A. approximately 3 billion New Taiwanese dollars

   B. more than 3 billion New Taiwanese dollars

   C. nearly 3 billion New Taiwanese dollars

5. Which Japanese company dominated the instant noodles market in Taiwan?

   A. Ve Wong Company

   B. the International Food Company

   C. the King Car Group

Situation 5

1. How long does an MRI exam typically last?

   A. from 5 to 15 minutes

   B. from 20 to 45 minutes

   C. from 45 minutes to an hour

2. What does the magnetic chamber include?

   A. an intercom system

   B. a contrast agentj

   C. axial, sagittal and coronal observations

3. What is the average income in Taiwan?

   A. $US 13,000

   B. $US 18,000

   C. $US 19,000

4. Why have Taiwanese become more health conscious and recreation-oriented?

   A. a strong economy

   B. advanced medical procedures

   C. improved living standards

5. What are patients free to do?

   A. consult with the attending physician or medical technologists

   B. have their relatives accompany them

   C. lie in a supine position and remain still

Situation 6

1. What does tomotherapy allow physicians to do?

   A. increase cure rates for cancer patients

   B. offer precise planning through using a treatment planning optimizer

   C. verify treatment volumes in advance

2. How is the dose delivery for all patients optimized?

    A. by using a treatment planning optimizer that is easier to use than conventional treatment planning systems

    B. by concentrating the radiation on the tumor and depositing less radiation in surrounding healthy tissue

    C. by delivering helical tomotherapy to targets while minimizing damage to healthy tissue

3. What can this therapeutic treatment system help hospital oncology departments to do?

    A. ensure precise positioning through using a unique verification CT

    B. provide enhanced medical care for cancer patients

    C. offer the most advanced radiation delivery system available

4. Where and when was tomotherapy pioneered?

    A. at the University of Wisconsin-Madison ten years ago

    B. at the University of Michigan ten years ago

    C. at the University of Wisconsin-Madison fifteen years ago

5. How can tomotherapy ensure precise delivery of the prescribed dosage to the intended area?

    A. owing to its ability to combine a treatment planning optimizer, a linear accelerator, computed tomography (CT) and a complex intensity modulation radiation therapy (IMRT)

    B. owing to its ability to concentrate the radiation on the tumor and depositing less radiation in surrounding healthy tissue

    C. owing to its ability to combine complex IMRT with spiral delivery

O Reading Comprehension II

Pick the work or expression whose meaning is closest to the meaning of the underlined word or expression in the following passages.

Situation 4

1. Fast food items are <u>immensely</u> popular among Taiwanese, with instant noodles being no exception.

   A. superficially

   B. expansively

   C. intangibly

2. Delicious, convenient, <u>inexpensive</u> and healthy, instant noodles have been a staple food item among Taiwanese for more than four decades.

   A. exorbitant

   B. economical

   C. extortionate

3. When production of instant noodles in Taiwan began in 1967, the International Food Company from Japan initially <u>dominated</u> the market.

   A. domineered

   B. abdicated

   C. renounced

4. However, after Wei Lih Food Manufacturers established a food processing plant in Changhua in 1970, local producers began to gradually <u>erode</u> Japan's market dominance.

   A. substantiate

   B. invigorate

   C. corrode

5. Taiwanese manufacturers initially imitated Japanese products, but eventually they began making adjustments to <u>appeal</u> to local tastes, such as adding chicken essence to instant noodles and enclosing seasoning packets that included salt, monosodium glutamate, pepper and other flavorings.

   A. repel

   B. entice

   C. disperse

6. With other local enterprises entering the market, including Ve Wong Company, Uni-President Enterprises, Vedan Enterprise Corporation and even the King Car Group, local production of instant noodles gradually <u>matured</u>, and local products gradually captured the dominant market share.

   A. seasoned

   B. retroacted

   C. receded

7. In 2002, after 37 years in business, Taiwanese manufacturers of instant noodles achieved <u>revenues</u> of approximately 3 billion New Taiwanese dollars.

   A. profits

   B. shortfall

   C. arrears

8. According to the 2003 Integrated Consumer <u>Tendency</u> (ICT) report on Taiwanese consumer trends, 15-29 year olds are the biggest consumers of instant noodles.

   A. Melancholy

   B. Despondency

   C. Inclination

9. Increasing market demand for diet food products has led to <u>innovations</u> in instant noodles.

   A. atrophy

B. modernization

C. degeneration

10. Furthermore, Taiwan's recent <u>entry</u> into the World Trade Organization has created opportunities for technology cooperation aimed at better satisfying consumer tastes, enhancing production management practices and improving after-sales service.

A. ingress

B. emanation

C. departure

11. Given the above trends, local manufacturers of instant noodles <u>face</u> new opportunities and challenges.

A. omit

B. overlook

C. contemplate

Situation 5

1. The Taiwanese economy has grown strong during the past decade, and the <u>average</u> income has now reached $US 13,000.

A. unorthodox

B. unprecedented

C. standard

2. Improved living standards have made Taiwanese more health conscious and <u>recreation</u>-oriented.

A. rejuvenation

B. impediment

C. oppression

3. Although most employees undergo a routine physical examination annually, including blood tests, chest x-ray examinations and heart-lung function testing, such examinations do not accurately <u>reflect</u> the current condition of patients.

A. sidetrack

B. resonate

C. divert

4. Most individuals <u>pay</u> extra for an MRI examination when undergoing their routine physical examination.

A. vend

B. remit

C. auction off

5. No longer <u>restricted</u> simply to identifying lesions, MRI examinations have become an effective means of determining the current status of human organs and vessels.

A. confined

B. approachable

C. genial

6. One of the advantages of an MRI exam is that no prior <u>preparations</u> are necessary.

A. oversights

B. omissions

C. provisions

7. Patients can eat normally, <u>continue</u> with their normal daily routines and continue taking any prescribed medications.

A. retrench

B. commence

C. curtail

8. Typically lasting from 20 to 45 minutes, depending on the information required by the physician, the procedure simply requires the patient to lie in a supine position and remain still.

   A. standing

   B. prostrate

   C. upright

9. Patients can be <u>accompanied</u> by relatives in the scan room, and are closely supervised by medical technologists.

   A. escorted

   B. abandoned

   C. cast off

10. Additionally, the magnetic chamber includes an intercom system should the patient <u>require</u>.

    A. spurn

    B. abstain from

    C. insist upon

11. A contrast agent may be administered to enhance the study, but no <u>precautions</u> are necessary.

    A. vulnerabilities

    B. perils

    C. safeguards

12. Patients are free to <u>consult</u> with the attending physician or medical technologists to discuss any concerns.

    A. quell

    B. deliberate

    C. subdue

13. Importantly, the examination involves no radiation, with data <u>acquired</u> via other means, which include axial, sagittal and coronal observations.

A. procured

B. forfeited

C. vanished

14. Hospitals increasingly realize the potential of <u>comprehensive</u> physical examinations for generating revenue, thus reducing pressures on the already strained national health insurance system.

A. hampered

B. hemmed in

C. inclusive

Situation 6

1. As a novel radiation therapy and planning system that can increase cure rates for cancer patients, tomotherapy offers the most advanced radiation delivery system available through its enhanced dose modulation and <u>accurate</u> targeting of specific locations.

A. erroneous

B. precise

C. unsound

2. As a novel radiation therapy and planning system that can increase cure rates for cancer patients, tomotherapy offers the most advanced radiation delivery system available through its enhanced dose modulation and accurate targeting of <u>specific</u> locations.

A. precise

B. vague

C. ambiguous

3. Tomotherapy allows physicians to <u>verify</u> treatment volumes in advance through 3D imagery via TomoImage scanning, ensuring that treatment fits a therapeutic strategy.

A. brush off

B. gloss over

C. substantiate

4. Additionally, this system delivers helical tomotherapy to targets while minimizing damage to healthy tissue, thus <u>optimizing</u> dose delivery for all patients.

A. maximizing

B. diminishing

C. depreciating

5. <u>Pioneered</u> by Professor Thomas Rockwell Mackie and the mathematician and software engineer Paul J. Reckwerdt at the University of Wisconsin-Madison ten years ago, tomotherapy combines a treatment planning optimizer, a linear accelerator, computed tomography (CT) and a complex intensity modulation radiation therapy (IMRT).

A. Sabotaged

B. Undermined

C. Created

6. Pioneered by Professor Thomas Rockwell Mackie and the mathematician and software engineer Paul J. Reckwerdt at the University of Wisconsin-Madison ten years ago, tomotherapy combines a treatment planning optimizer, a linear accelerator, computed tomography (CT) and a complex <u>intensity</u> modulation radiation therapy (IMRT).

A. potency

B. infirmity

C. decrepitude

7. Among its <u>unique</u> features, tomotherapy offers precise planning through using a treatment planning optimizer that is easier to use than conventional treatment planning systems.

    A. universal

    B. peerless

    C. orthodox

8. Moreover, tomotherapy ensures precise positioning through using a unique verification CT to confirm the tumor position before each treatment, enabling precise <u>delivery</u> of the radiation dosage.

    A. allotment

    B. retraction

    C. recantation

9. Furthermore, tomotherapy also ensures precise delivery of the prescribed dosage to the <u>intended</u> area owing to its ability to combine complex IMRT with spiral delivery, thus concentrating the radiation on the tumor and depositing less radiation in surrounding healthy tissue.

    A. involuntary

    B. inadvertent

    C. contrived

10. In sum, this therapeutic treatment system is widely <u>anticipated</u> to be adopted among hospital oncology departments to provide enhanced medical care for cancer patients.

    A. miscalculated

    B. forecasted

    C. undervalued

Unit Three

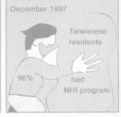
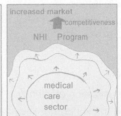

Describing a Project
for Developing a Product or a Service

專案描述

1. Provide the rationale for taking on the project.
   專案的背景
2. Present market data that justifies the project's feasibility.
   市場的訊息
3. Spell out the immediate and long term goals of the project.
   專案的目標
4. Describe the project's distinguishing features, while pointing out the strategy employed to successfully complete it.
   專案的重要特色：專案的策略
5. Conclude the presentation by pointing out the anticipated merits of the project and the positive impact that it will have.
   專案的動向

Vocabulary and related expressions

considerable portion of	相當多的部分
lags behind	延遲
large trade imbalance	高度的貿易不均衡狀態
satisfy domestic demand	滿足國內需求
state-of-the-art	（科技、機電等產品）最先進的，最高級的
fashionable trends	流行潮流
raise brand awareness	提高品牌知名度
appeal to	吸引
initiated	開始實施
mandatory	命令的
resource utilization efficiency	資源使用的效率性
increased market competitiveness	增加市場競爭力
operational efficiency	經營上效率
newly emerging market	新增市場
electronic journals	電子期刊
slow convergence	遲緩的聚合
conventional software	傳統的軟體
a promising source of revenue	有希望的收益來源
many unexplored areas	許多未探測的區域
a mature technology	成熟的科技
preliminary results	初步的結果
increasing market demand	日漸增加的市場需求
strong government backing	有力的政府支援
the latest technological applications	最新的科技應用
effective market strategies	有效的市場策略
promotion strategies	（商品等的）促銷策略

Situation 1

Situation 2

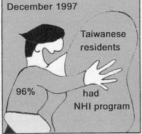
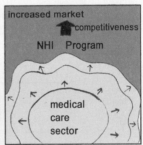

Situation 3

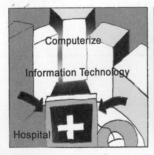
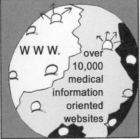

A Write down the key points of the situations on the preceding page, while the instructor reads aloud the script from the Answer Key. Alternatively, students can listen to the script online at www.chineseowl.idv. tw

Situation 1

_____

_____

Situation 2

_____

_____

_____

Situation 3

_____

_____

_____

B Oral practice I

Based on the three situations in this unit, write three questions beginning with **Why**, and answer them. The questions do not need to come directly from these situations.

Examples

Why does a large trade imbalance exist between Taiwan and Korea in on-line gaming? Because Taiwan lags behind Korea in terms of development

Why has the Taiwanese government added on-line game to the priority list of growth areas for national development?

Owing to its enormous potential

1. _____

   _____

2. _____

   _____

3. _____

   _____

C Based on the three situations in this unit, write three questions beginning with ***What***, and answer them. The questions do not need to come directly from these situations.

Examples

What percentage of residents of Taiwan was covered by 13 different health insurance schemes before the Taiwanese government initiated the National Health Insurance Program in 1995

Nearly 60%

What did the Council for Economic Planning and Development establish a planning committee in 1988 to do?

Develop a single mandatory and universal health insurance program

1. _____

_____

2. _____

_____

3. _____

_____

D Based on the three situations in this unit, write three questions beginning with *How*, and answer them. The questions do not need to come directly from these situations.

Examples

How were Taiwanese hospitals impelled to computerize their operations?

With the establishment of the National Health Insurance scheme in 1995

How can these duplicate IC cards be used?

For medical payments

1. _____

_____

2. _____

_____

3. _____

_____

E Write questions that match the answers provided.

1. _____

_____

_____

Eight years ago

2. _____

_____

_____

Many information technology firms to this newly emerging market

3. _____

_____

_____

2600 clinics and 47 hospitals islandwide

F Listening Comprehension I

Situation 1

1. Which country does Taiwan lag behind in terms of development of on-line gaming?

   A. Hong Kong

   B. China

   C. Korea

2. How has the Taiwanese government encouraged the development of on-line gaming?

   A. by adding this sector to the priority list of growth areas for national development

   B. by promoting the connection between the new game and current comic book series

   C. by emphasizing design, manufacturing and testing

3. What is the key market sector for on-line gaming?

   A. 7-12 year olds

   B. 13-26 year olds

   C. 30-45 year olds

4. How do current government policies aim to satisfy domestic demand for on-line games?

   A. by emphasizing design, manufacturing and testing

   B. by replacing imported Korean products with innovative local products

   C. by representing a considerable portion of the game sector

5. Why are new on-line games promoted together with current comic book series, movies and television programs?

   A. to raise brand awareness

B. to emphasize design, manufacturing and testing

C. to appeal to Chinese consumers

Situation 2

1. How could the NHI Bureau help control hospital health care costs and resource utilization efficiency?

   A. by expanding the size and scope of the medical care sector

   B. by creating incentives for continuously improving operational efficiency

   C. by implementing a Global Budget payment system and introducing payment on a per case basis rather than a per-visit basis

2. What was the first reason why the NHI Bureau was established?

   A. to improve medical care quality and increase competition among healthcare providers

   B. to enable the economic provision of high quality health care services

   C. to ensure that all Taiwanese residents received insurance coverage

3. What percentage of all Taiwanese residents had NHI program coverage as of December 1997?

   A. around 95%

   B. around 96%

   C. nearly 100%

4. How does the NHI system differ from the previous one?

   A. Patients can freely select where they receive medical care.

   B. 40% of the population is left to pay for treatment entirely on their own.

   C. Incentives are available for continuously improving operational efficiency.

5. What aspect of the medical care sector has the NHI program expanded?

   A. its size and scope

   B. its financial profits

C. its geographical reach

Situation 3

1. What percentage of all hospitals in Taiwan outsourced their information system needs eight years ago?

   A. more than 10%

   B. around 10%

   C. less than 10%

2. What impelled Taiwanese hospitals to computerize their operations?

   A. the establishment of the National Health Insurance scheme in 1995

   B. the trend of credit cards becoming a standard payment method

   C. the emergence of many information technology firms

3. How many clinics comprise the customer base for the information technology sector?

   A. 2500

   B. 2600

   C. 2700

4. How many medical information-oriented websites are there globally?

   A. over 100,000

   B. over 1,000

   C. over 10,000

5. Why did Real-Sun design IC-based duplicate cards for medical treatment?

   A. concerns regarding how to provide access to electronic journals

   B. concerns regarding how to protect client information in an IC card format

   C. governmental policy aimed at upgrading the computerized capabilities of hospitals

G Reading Comprehension I
Pick the work or expression whose meaning is closest to the meaning of the underlined word or expression in the following passages.

Situation 1

1. Despite representing a considerable portion of the game sector, on-line gaming in Taiwan <u>lags</u> behind Korea in terms of development, explaining the large trade imbalance between the two countries in this industry.

    A. accelerates

    B. wanes

    C. expedites

2. Despite representing a <u>considerable</u> portion of the game sector, on-line gaming in Taiwan lags behind Korea in terms of development, explaining the large trade imbalance between the two countries in this industry.

    A. inapplicable

    B. irrelevant

    C. appreciable

3. Owing to the <u>enormous</u> potential of on-line gaming, the Taiwanese government has added this sector to the priority list of growth areas for national development.

    A. piddling

    B. immense

    C. dinky

4. Current government policies aim not only to satisfy <u>domestic</u> demand for on-line games by replacing imported Korean products with innovative local products, but also to develop state-of-the-art globally competitive online games by emphasizing design, manufacturing and testing.

A. indigenous

B. international

C. global

5. Current government policies aim not only to satisfy domestic <u>demand</u> for on-line games by replacing imported Korean products with innovative local products, but also to develop state-of-the-art globally competitive online games by emphasizing design, manufacturing and testing.

A. pretension

B. deception

C. mandate

6. Additionally, the government is <u>encouraging</u> two novel promotional strategies.

A. endorsing

B. deterring

C. squelching

7. First, <u>fashionable</u> trends and comic characters that appeal to the key market sector, 13-26 year olds, are adopted in game design and promotion.

A. modish

B. shunned

C. scorned

8. Second, new games are promoted together with current comic book series, movies and television programs to <u>raise</u> brand awareness.

A. downplay

B. elevate

C. subdue

9. Although this strategy requires considerable time investment to promote the connection between the new game and current comic book series, movies and television programs, it offers a means of design and marketing on-line games that

appeal to Chinese consumers.

A. repulse

B. annoy

C. allure

10. Although this strategy requires considerable time investment to promote the connection between the new game and current comic book series, movies and television programs, it offers a means of design and marketing on-line games that appeal to Chinese consumers.

A. schism

B. fissure

C. juncture

Situation 2

1. Before the Taiwanese government initiated the National Health Insurance Program in 1995, 13 different health insurance schemes covered nearly 60% of residents of Taiwan, with the remaining population left to pay for treatment entirely on their own.

A. impeded

B. desisted

C. inaugurated

2. To effectively address this problem, the Council for Economic Planning and Development established a planning committee in 1988 to develop a single mandatory and universal health insurance program.

A. volitional

B. obligatory

C. voluntary

3. The program took effect in March 1995 after the Legislative Yuan passed the

National Health Insurance (NHI) Act.

A. renounced

B. rebuffed

C. approved

4. The NHI Bureau was established to <u>achieve</u> three goals.

A. effectuate

B. abort

C. flounder

5. First, the NHI Bureau sought to <u>ensure</u> that all Taiwanese residents received insurance coverage.

A. trammel

B. subjugate

C. confirm

6. According to the Department of Health of Taiwan, <u>around</u> 96% of all Taiwanese residents had NHI program coverage as of December 1997.

A. an excessive of

B. approximately

C. more than

7. Second, the NHI Bureau should improve medical care quality and increase <u>competition</u> among healthcare providers.

A. collaboration

B. contention

C. cooperation

8. Unlike under the previous system, under NHI patients can <u>freely</u> select where they receive medical care.

A. casually

B. ceremonially

C. rigidly

9. Third, by <u>implementing</u> a Global Budget payment system and introducing payment on a per case basis rather than a per-visit basis, the NHI Bureau could help control hospital health care costs and resource utilization efficiency.

A. discarding

B. jettisoning

C. promulgating

10. This approach <u>enabled</u> the economic provision of high quality health care services.

A. mutilate

B. capacitated

C. mangle

11. In sum, the NHI program has expanded the size and scope of the medical care sector, increased market competitiveness, and created <u>incentives</u> for continuously improving operational efficiency.

A. impetuses

B. impediments

C. stumbling blocks

Situation 3

1. When Real-Sun Information Technology Company <u>entered</u> the medical services sector in Taiwan eight years ago, less than 10% of all hospitals outsourced their information system needs.

A. vacated

B. penetrated

C. retreated from

2. However, the establishment of the National Health Insurance scheme in 1995 impelled Taiwanese hospitals to computerize their operations.

A. contravened

B. repressed

C. induced

3. Moreover, governmental policy aimed at upgrading the computerized capabilities of hospitals attracted many information technology firms to this   newly emerging market.

A. emanating

B. reverting

C. recoiled

4. Consequently, nearly 80 information technology companies invested in the medical information sector, and Internet companies followed them into the market, representing an initial investment of nearly $NT 20,000,000.

A. rudimentary

B. conclusive

C. terminating

5. Following several company mergers and considerable investment, Taiwan has around 20 domestic information technology firms involved in the medical sector.

A. scissions

B. fragmentations

C. conglomerations

6. The customer base for this sector comprises 2600 clinics and 47 hospitals island-wide.

A. precludes

B. consists of

C. disallows

7. Real-Sun Information Technology Company has recently <u>merged</u> its personnel and resources with another medical software firm to increase its market share.

A. demarcated

B. severed

C. amalgamated

8. As domestic medical health networks continue to <u>evolve</u> in this newly emerging market, 40 local firms have been established, while globally there are over 10,000 medical information-oriented websites.

A. wind up

B. unfold

C. terminate

9. Several websites provide online research capabilities, providing <u>access</u> to electronic journals, disease-related information for diagnostic purposes, on-line queries, examinations, automatic registration functions, personal health tips and online ordering capabilities.

A. ingress

B. impediment

C. obstruction

10. Meanwhile, all Internet-based medical information companies can handle kinesiology and medicine classification-related <u>queries</u>.

A. submissions

B. propositions

C. inquiries

11. Concerns regarding how to <u>protect</u> client information in an IC card format make extracting information from IC card contents for commercial purposes extremely difficult.

A. disclose

B. safeguard

C. unmask

12. Real-Sun has therefore designed IC-based <u>duplicate</u> cards for medical treatment.

A. facsimile

B. bona fide

C. authentic

13. As credit cards have become a standard payment method, these duplicate IC cards can be used for medical payments, <u>enabling</u> thousands of clinics to adopt a uniform method of payment processing.

A. averting

B. precluding

C. facilitating

H Common elements in describing a project for developing a product or service專案描述include the following contents:

1. Provide the rationale for taking on the project.
專案的背景

· Petroleum energy crises have alerted the general popularity over the scarcity of non-renewable resources and the necessity to develop alternative renewable energy sources such as solar power.

· The fiercely competitive management environment has necessitated that enterprises initiate many innovative plans, including organizational restructuring or personnel reductions. Likewise, outsourcing is a major industrial trend aimed at ensuring corporate survival.

· Why are so many small-scale healthcare institutes in Taiwan actively engaged in projects aimed at elevating the quality of long term care? First, as Taiwan's rapidly growing elderly population reflects global aging trends, individuals without professional training account for 69% of all healthcare providers of the disabled elderly.

2. Present market data that justifies the project's feasibility.
市場的訊息

· According to a recent market survey, although only around 20% of all disabled elderly in Taiwan receive institutional-based care, the market demand for institutional-based care is 30% (Department of Health, 1997). The 10% difference is equivalent to a market scale of at least 18,000 individuals.

· With consumer demand expected to grow continuously, according to expert forecasts, solar energy use will increase at a rate of 33% the future, with its output value targeted at $US 1,400,000,000.

· As a body mass index (BMI) value of 24-27 is considered fat in Taiwan, an increasing number of people fall within this range, putting them at a 60% risk of diabetes mellitus. Consequently, the market for slimming products has exploded in recent years, exceeding $NT 40,000,000,000 in Taiwan last year.

3. Spell out the immediate and long term goals of the project.
　專案的目標

· More than an alternative to petroleum, as an unlimited power source that does not extract an environmental cost, solar energy represents a growing environmental consciousness aimed at protecting the world's natural resources.

· The government should encourage this initiative in two ways. Legislation governing health food standards should be implemented, as is already done so globally. Second, efforts should be directed towards strengthening the production capacity of the health food sector locally because the mostly imported health food products currently available do not necessarily suit local tastes and consumer demand.

· To satisfy a broad spectrum of consumer tastes and demand for certain flavors, producers will develop an extensive product line of homemade style cookies and other snack varieties, including bread, cake and candy. Substantial investment must also be made in machinery and preservative containers to ensure that the cookies maintain their freshness and avoid the use of unhealthy additives that contribute to obesity.

4. Describe the project's distinguishing features, while pointing out the strategy employed to successfully complete it.

專案的重要特色：專案的策略

· Unlike non-renewable sources, solar energy is unique in that tremendous amounts of capital are not necessary to develop its potential and make it applicable for human needs, further eliminating the need to pose an environmental burden.

· Under the BOT scheme, the Taiwanese government initially holds the property rights of an infrastructure project and then transfers them to the private sector. The private sector is then responsible for investing in the project and implementing it for an agreed upon period. Once the period expires, the private sector transfers the project assets to the government with or without compensation.

· Given consolidation of medical personnel and subsequent increased workload in the healthcare sector, hospital administrative strive to effectively manage human resources as flexibly as possible. As clinical management adopts various outsourcing strategies to alleviate the above predicament, whether outsourcing will extend to nursing personnel, which account for the largest portion of the hospital staff, to control expenditures and better allocate human resources remains to be seen.

5. Conclude the presentation by pointing out the anticipated merits of the project and the positive impact that it will have.

專案的動向

· While successful development of Taiwan's hi-tech industry is reflected in the Hsinchu Science-based Industrial Park and the Tainan Science Park, the new transportation network will further accelerate access and technological progress between the northern and southern parts of the island.

· County government officials hope that the newly established THSR stations will spur tourist and economic growth.

· Its projected completion in 2006 will significantly transform the island's transportation market sector.

Describe a project for developing a product or a service.

_____

_____

_____

_____

_____

_____

_____

_____

_____

_____

Look at the following examples of how to describe a project for developing a product or a service.

Petroleum energy crises have alerted the general popularity over the scarcity of non-renewable resources and the necessity to develop alternative renewable energy sources such as solar power. The Industrial Technology Research Institute initiated efforts to develop solar energy-based technologies in Taiwan in 1982. Although incapable of replacing petroleum use entirely, the seemingly endless potential of solar energy has been extensively developed worldwide, as fueled by a growing environmental consciousness. For instance, hot-water machines powered by solar energy are widely available commercially, covering an area of 885,500 square meters. With consumer demand expected to grow continuously, according to expert forecasts, solar energy use will increase at a rate of 33% the future, with its output value targeted at $US 1,400,000,000. More than an alternative to petroleum, as an unlimited power source that does not extract an environmental cost, solar energy represents a growing environmental consciousness aimed at protecting the world's natural resources. As is widely anticipated, solar energy has the potential for many products and appliances, including automobiles and notebook computers. Unlike non-renewable sources, solar energy is unique in that tremendous amounts of capital are not necessary to develop its potential and make it applicable for human needs, further eliminating the need to pose an environmental burden. A notable example of its success is the solar energy powered hot-water machines in Taiwan, which have saved 74 million liters of water. As scientists continuously explore how to integrate solar energy with a diverse array of machinery, solar energy still has several limitations that must be resolved, such as the inability to use solar energy-powered devices for extended periods or in overcast weather. Still, efforts to address these limitations are aggressively underway, with optimism for even greater use in the near future.

Government-sponsored Build-Operate-Transfer (BOT) projects encourage the private sector to participate in infrastructure projects. Under the BOT scheme, the Taiwanese government initially holds the property rights of an infrastructure project and then transfers them to the private sector. The private sector is then responsible for investing in the project and implementing it for an agreed upon period. Once the period expires, the private sector transfers the project assets to the government with or without compensation. For instance, a major transportation construction project such as the Taiwan High Speed Rail (THSR) project is an effective means by which the government promotes regional development. Promoted for its high speed, punctuality and capacity, THSR will greatly facilitate accessibility between each region in Taiwan. County government officials hope that the newly established THSR stations will spur tourist and economic growth. Its projected completion in 2006 will significantly transform the island's transportation market sector. Projected expenses on the BOT-sponsored THSR rail transportation system will reach 500 billion New Taiwanese dollars, making it the most expensive of its kind worldwide. Furthermore, the slab track-based rail transportation system will enable speeds exceeding 300 km per hour, making it possible to reach Kaohsiung from Taipei, i.e., a distance of 345 km, in only 90 minutes. Such capability will truly make Taiwan a metropolitan area island wide. In contrast with other transportation modes, the high speed rail is more rapid, punctual and safe. While successful development of Taiwan's hi-tech industry is reflected in the Hsinchu Science-based Industrial Park and the Tainan Science Park, the new transportation network will further accelerate access and technological progress between the northern and southern parts of the island. Since construction began in 1999, THSR has raised the optimism of many Taiwanese residents with respect to sustained economic growth Still, effective marketing strategies must be adopted to lure customers away from their cars as a viable option to the already congested national freeway infrastructure.

Commonly consumed foodstuffs found in convenience stores, supermarkets and restaurants contain an increasing amount of potentially harmful additives and preservatives. As home cooked meals become less frequent owing to the hectic lifestyles of working couples, food that is greasy and high in sodium is consumed outside of the home, contributing to obesity and poor health. As an alternative, the health food market in Taiwan has greatly expanded in recent years, with nearly 70% of those products imported from abroad. This figure represents a large gap that local health food producers could fill by offering products that appeal to local tastes. Local research efforts should be directed towards supplying a sufficient quantity and diversity of health food that complies with stringent nutrition standards. The government should encourage this initiative in two ways.? Legislation governing health food standards should be implemented, as is already done so globally. For instance, China, Japan and the United States have adopted legislation to spur market growth of the health food sector. In line with these trends, the Taiwanese government could hopefully launch measures to standardize management procedure in the health food sector, thus paving the way for continuous growth of this newly emerging market. Second, efforts should be directed towards strengthening the production capacity of the health food sector locally because the mostly imported health food products currently available do not necessarily suit local tastes and consumer demand. Despite the considerable time and capital investment required to successfully implement the above two strategies, doing so would not only encourage local food producers to opt for healthy alternatives, but also fill in the gap that is largely controlled by overseas producers.

Situation 4

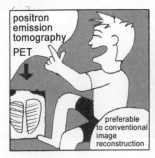
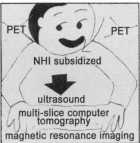
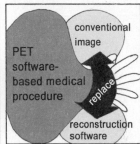

Situation 5

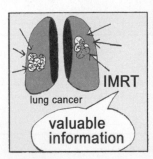
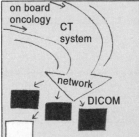
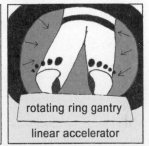

Situation 6

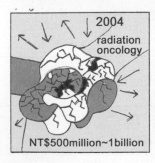
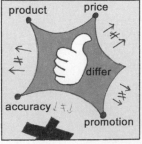
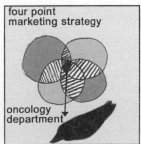

I Write down the key points of the situations on the preceding page, while the instructor reads aloud the script from the Answer Key. Alternatively, students can listen to the script online at www.chineseowl.idv. tw

Situation 4

_____

_____

_____

Situation 5

_____

_____

_____

Situation 6

_____

_____

_____

J Oral practice II

Based on the three situations in this unit, write three questions beginning with **What**, and answer them. The questions do not need to come directly from these situations.

Examples

What does conventional image reconstruction require?

Filtered back projection software and maximum likelihood expectation maximization software that often produces poor quality images owing to the limited number of photons and the slow convergence of original image data

What has been widely adopted in experimental investigations using PET, and involves a fast algorithm enhancing the slow convergence of conventional software? An image reconstruction method based on ordered subset expectation maximization software

1. _____

   _____

2. _____

   _____

3. _____

   _____

K Based on the three situations in this unit, write three questions beginning with **Why**, and answer them. The questions do not need to come directly from these situations.

Examples

Why is tomotherapy an excellent therapy for lung cancer patients?

It not only enables precise image-guided intensity modulated radiotherapy, but also provides valuable information regarding tumor changes during radiotherapy.

Why are physicians able to determine whether a tumor is shrinking?

TomoImage scans are taken before each treatment.

1. _____

_____

2. _____

_____

3. _____

_____

L Based on the three situations in this unit, write three questions beginning with *How*, and answer them. The questions do not need to come directly from these situations.

Examples

How is it evident that medical technology has rapidly evolved recently?

By the increasing number of technologies available to radiation oncology departments, for example radiation pharmaceuticals and linear accelerators

How is the increasing market demand for advanced cancer therapeutic treatment strategies evident?

Revenues from the radiation oncology sector in Taiwan ranged between $NT 500 million to 1 billion dollars in 2004.

1. _____

_____

2. _____

_____

3. _____

_____

M Write questions that match the answers provided.

_____

_____

The increasing number of technologies available to radiation oncology departments

_____

_____

Between $NT 500 million to 1 billion dollars

_____

_____

Because of the large market potential and strong government backing

N Listening Comprehension II

Situation 4

1. When did the mathematicians Hudson and Larkin pioneer an image reconstruction method based on ordered subset expectation maximization software?

   A. in 1992

   B. in 1993

   C. in 1994

2. Which group can choose from other less advanced but still high quality procedures subsidized by NHI?

   A. patients that are unwilling to cover the expense of this PET examination themselves

   B. hospitals that are unwilling to cover the expense of this PET examination

   C. governmental authorities that are unwilling to cover the expense of this PET examination

3. What can this PET software-based medical procedure eventually replace?

   A. conventional algorithms that enhance the slow convergence of conventional software

   B. conventional computer image reconstruction software

   C. conventional positron emission tomography procedures

4. How many patients has the PET Center of Shin Kong Wu Ho-Su Memorial Hospital served over the past two years?

   A. over 4,000 patients

   B. nearly 4,000 patients

   C. around 4,000 patients

5. Why are several hospitals developing marketing strategies?

   A. to generate revenues owing to budgetary constraints

B. to compensate for unsubsidized services

C. to make patients aware of such high quality medical treatments

Situation 5

1. What does tomotherapy provide valuable information regarding?

    A. many unexplored areas in radiation oncology

    B. tumor changes during radiotherapy

    C. the size, shape and intensity of the radiation beam

2. How many patented technologies does Tomotherapy Incorporated hold in this area?

    A. 65

    B. 70

    C. 75

3. What delivers a fan beam of photon radiation as the ring turns?

    A. a linear accelerator

    B. a multileaf collimator

    C. TomoImage

4. Who does the Cancer Center at NTU plan to collaborate with?

    A. lung cancer patients

    B. radiation oncologists

    C. Professor Thomas Rockwell Mackie and Paul J. Reckwerdt

5. What does the precision of tomotherapy offer?

    A. The ability to determine whether a tumor is shrinking

    B. therapeutic potential for cancer patients ineligible for radiotherapy

    C. available resources to combat cancer

Situation 6

1. How much in revenues did the radiation oncology sector in Taiwan generate in 2004?

A. between $NT 150 million to $400 billion dollars

B. between $NT 500 million to 1 billion dollars

C. over 1 billion dollars

2. Why have hospitals with radiation oncology departments expressed strong interest in adopting the latest technological applications?

A. Given the large market potential and strong government backing

B. Given a four-point based market strategy that can help clinical radiation oncology departments equip management professionals with appropriate and efficient marketing policies

C. Given an intensely competitive global market that emphasizes state-of-the-art medical instrumentation and professionalism

3. What does the four-point based marketing strategy focus on?

A. state-of-the-art medical instrumentation and professionalism

B. radiation pharmaceuticals and linear accelerators

C. technology differentiation

4. What do instrumentation companies strive to do?

A. prepare radiation oncology departments in Taiwan for an intensely competitive global market

B. efficiently use available resources and identify effective market strategies

C. help clinical radiation oncology departments equip management professionals with appropriate and efficient marketing policies

5. What is there an increasing market demand for?

A. promotion strategies with unique features that differentiate them from other products

B. an increasing number of technologies available to radiation oncology departments

C. advanced cancer therapeutic treatment strategies

O Reading Comprehension II

Pick the work or expression whose meaning is closest to the meaning of the underlined word or expression in the following passages.

Situation 4

1. As a non-invasive form of nuclear medicine, positron emission tomography (PET) is preferable to <u>conventional</u> image reconstruction, which requires filtered back projection software and maximum likelihood expectation maximization software that often produces poor quality images owing to the limited number of photons and the slow convergence of original image data.

   A. offbeat

   B. eccentric

   C. orthodox

2. As a non-invasive form of nuclear medicine, positron emission tomography (PET) is preferable to conventional image reconstruction, which requires filtered back projection software and maximum likelihood expectation maximization software that often produces poor quality images owing to the limited number of photons and the slow <u>convergence</u> of original image data.

   A. scission

   B. segmentation

   C. confluence

3. <u>Pioneered</u> by the mathematicians Hudson and Larkin in 1994, an image reconstruction method based on ordered subset expectation maximization software has been widely adopted in experimental investigations using PET, and involves a fast algorithm enhancing the slow convergence of conventional software.

A. replicated

B. originated

C. simulated

4. Given the above <u>considerations</u>, hospitals must replace conventional software to produce high quality clinical images.

A. deliberations

B. oversight

C. disregard

5. Owing to budgetary <u>constraints</u>, National Health Insurance (NHI) will not cover the expense of this service, making this area a promising source of revenue for hospitals with the necessary technological capacity.

A. obstructions

B. contingency

C. opening

6. Patients that are <u>unwilling</u> to cover the expense of this PET examination themselves, can choose from other less advanced but still high quality procedures subsidized by NHI, such as ultrasound, multi-slice computer tomography or magnetic resonance imaging.

A. receptive

B. amenable

C. reluctant

7. A notable example of such unsubsidized services is the PET Center of Shin Kong Wu Ho-Su Memorial Hospital, which has served over 4,000 patients over the past two years, generating substantial revenues for the hospital, thus reducing its <u>reliance</u> on NHI for funding.

A. emancipation

B. dependence

C. self-determination

8. Each patient pays roughly $US 1,500 for this PET examination. Besides enabling the nuclear medicine departments in Taiwanese hospitals to fill a profitable market niche, this PET software-based medical procedure can <u>eventually</u> replace conventional computer image reconstruction software.

A. untimely

B. ultimately

C. inopportunely

9. Enhanced image reconstruction computer software based on maximizing ordered subset expectations is highly <u>promising</u> in clinical efforts to provide high-quality PET images.

A. auspicious

B. improbable

C. implausible

10. Several hospitals are currently developing marketing strategies to make patients <u>aware</u> of such high quality medical treatments.

A. unsuspecting

B. oblivious

C. cognizant

Situation 5

1. As an excellent therapy for lung cancer patients, tomotherapy not only enables <u>precise</u> image-guided intensity modulated radiotherapy, but also provides valuable information regarding tumor changes during radiotherapy.

A. obscure

B. explicit

C. dubious

2. Tomotherapy thus represents the future of image-guided IMRT in cancer treatment. The Cancer Center of National Taiwan University (NTU) utilizes all available resources to <u>combat</u> cancer.

A. confront

B. abdicate

C. relinquish

3. Numerous investigations have demonstrated the <u>potential</u> of tomotherapy in many unexplored areas in radiation oncology.

A. potency

B. ineptitude

C. feebleness

4. As a <u>compact</u>, cost effective and high precision radiation therapeutic treatment system, tomotherapy includes a primary beam-shield that reduces the operational costs associated with room shielding, an on-board oncology CT system, a rapid inverse planning system with a built-in optimizer, a full network with DICOM input, and a built-in patient scheduler.

A. leviathan

B. grandiose

C. miniaturized

5. Taking TomoImage scans before each treatment enables physicians to determine whether a tumor is <u>shrinking</u>.

A. expanding

B. amplifiying

C. recoiling

6. After four or five weeks of <u>treatment</u>, the treatment dosage can be decreased as the tumor size shrinks.

A. curative measure

B. malady

C. pathology

7. Known as slice therapy because it <u>derives</u> from tomography or cross-sectional imaging, the tomotherapy system resembles a computed tomography system: the patient lies on a bench that moves continuously through a rotating ring gantry.

A. emits

B. originates

C. expels

8. The gantry is <u>fitted</u> with a linear accelerator, which delivers a fan beam of photon radiation as the ring turns.

A. separated

B. conjoined

C. divided

9. The couch movement combined with the gantry rotation means that the radiation beam <u>spirals</u> around the patient.

A. twists

B. obstructs

C. blocks

10. Tomotherapy <u>delivers</u> intensity modulated radiotherapy (IMRT) using a multileaf collimator.?

A. impedes

B. obstructs

C. transmits

11. A recent advance in radiation treatment involves IMRT <u>altering</u> the size, shape and intensity of the radiation beam depending on tumor size, shape and location.

A. distorting

B. amending

C. blemishing

12. Tomotherapy achieves <u>better</u> cancer patient survival rates than other therapeutic treatment systems.

    A. enhanced

    B. regressing

    C. repetitive

13. The Cancer Center at NTU plans to <u>collaborate</u> with two innovators in this field: Professor Thomas Rockwell Mackie, a leading medical physicist, and Paul J. Reckwerdt, an accomplished mathematician and software engineer.

    A. struggle

    B. compete

    C. cooperate

14. Moreover, Tomotherapy Incorporated holds 70 patented technologies in this area, providing a valuable <u>source</u> of information for further understanding advanced applications of the tomotherapy system.

    A. destination

    B. origin

    C. arrival

15. Although some clinicians do not consider tomotherapy a <u>mature</u> technology, preliminary results for treating prostate and lung cancer are encouraging.

    A. worldly

    B. rudimentary

    C. gullible

16. The precision of tomotherapy offers therapeutic potential for cancer patients <u>ineligible</u> for radiotherapy.

    A. approved

    B. authorized

C. unqualified

Situation 6

1. Medical technology recently has rapidly <u>evolved</u>, as evidenced by the increasing number of technologies available to radiation oncology departments, for example radiation pharmaceuticals and linear accelerators.

   A. retroacted

   B. progressed

   C. relapsed

2. Given the increasing market demand for advanced cancer therapeutic treatment strategies, according to the National Science Council of the Republic of China, Taiwan, <u>revenues</u> from the radiation oncology sector in Taiwan ranged between $NT 500 million to 1 billion dollars in 2004.

   A. proceeds

   B. arrears

   C. shortfall

3. Given the large market potential and strong government <u>backing</u>, hospitals with radiation oncology departments have expressed strong interest in adopting the latest technological applications.

   A. prohibition

   B. resistance

   C. support

4. When creating a new model in this intensely competitive market, instrumentation companies <u>strive</u> to efficiently use available resources and identify effective market strategies.

   A. aspire

   B. hesitate

C. procrastinate

5. As an effective market strategy for understanding this highly <u>competitive</u> market, the four-point based market strategy can help clinical radiation oncology departments equip management professionals with appropriate and efficient marketing policies.

   A. stagnant

   B. emulous

   C. idle

6. Comprising product, price function, accuracy and promotion, the four-point based marketing strategy focuses on technology differentiation, in which hospital administrators <u>stress</u> how their product lines differ from those of other hospital centers.

   A. emphasize

   B. discourage

   C. sidetrack

7. The price <u>strategy</u> sets prices based on the prices of competing products.

   A. disavowal

   B. repudiation

   C. schema

8. Since radiation oncology strongly emphasizes <u>accuracy</u>, related technologies require precise instrumentation.

   A. rectitude

   B. fallaciousness

   C. misinterpretation

9. New products face an <u>intensely</u> competitive market and, thus, promotion strategies must stress the unique features that differentiate them from other products.

A. poignantly

B. trivially

C. inconsequentially

10. Finally, it is important to use the most <u>appropriate</u> agent for marketing purposes.

A. inequitable

B. inapt

C. congruous

11. In sum, the four-point based marketing strategy can help prepare radiation oncology departments in Taiwan for an intensely competitive global market that emphasizes <u>state-of-the-art</u> medical instrumentation and professionalism.

A. venerable

B. avant-garde

C. archaic

Unit Four

Introducing a Company or Organization

公司或組織介紹

1. Briefly overview the industry to which your organization belongs.
   簡要陳述組織所屬之產業的概況
2. Describe your organization's mission as well as highlight its historical development.
   組織的使命；組織的發展沿革
3. Introduce the organizational structure.
   組織的架構
4. Highlight recent technical accomplishments within the organization.
   組織最新的科技成就
5. Point towards the organization's future directions.
   結論（未來發展方向）

Vocabulary and related expressions

retail and distribution hub	零售及通路中心
revenue growth	總收入成長
relatively inexpensive labor	相對便宜的勞力
segmentation marketing strategy	市場區隔策略
new employee orientation	新進員工的適應（情況）介紹
ample skilled staff	大量有技能的員工
exceptional customer service	優秀的顧客服務
strong organizational framework	堅固的組織架構
ensure customer satisfaction	確保顧客滿意
logistics services	物流服務
fully complies with	完全的遵從
promulgating	公布／發表
communicable diseases	會傳染的疾病
outstanding research capabilities	優秀的研究能力
globally renowned	有全球性的聲譽的
pharmaceutical	配藥的
foster medical talent	培養醫學天資
yield optimal research results	產出最佳的研究結果
successfully commercializing	成功的使商品化
constant technological innovation	接連不斷的科技創新
accreditation	鑑定
integrating	整合
continuously upgrading business operations	持續地提升商業運作
clinical trial facility	醫學臨床的試驗技能／設備

Situation 1

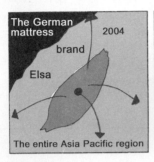
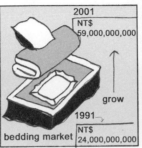
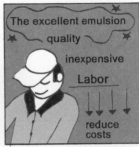

Situation 2

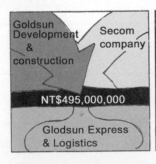

Situation 3

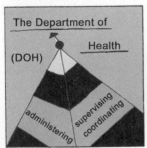
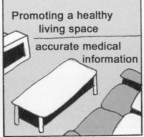

A Write down the key points of the situations on the preceding page, while the instructor reads aloud the script from the Answer Key. Alternatively, students can listen to the script online at www.chineseowl.idv. tw

Situation 1

_____

_____

Situation 2

_____

_____

_____

Situation 3

_____

_____

_____

B Oral practice I
Based on the three situations in this unit, write three questions beginning with **Why**, and answer them. The questions do not need to come directly from these situations.

Examples

Why has Elsa been able to reduce overhead costs and pass these savings on to consumers in the form of competitive retail prices?

Because of the excellent emulsion quality and relatively inexpensive labor in Malaysia

Why did Elsa select Taiwan as its retail and distribution hub in the Asia Pacific region?

Owing to careful consideration and extensive marketing research

1. _____

_____

2. _____

_____

3. _____

_____

C Based on the three situations in this unit, write three questions beginning with *When*, and answer them. The questions do not need to come directly from these situations.

Examples

How was Goldsun Express & Logistics (GE&L) Company (formerly Kuo Hsing Transport) established?

As a joint investment by Taiwan Secom Company and Goldsun Development & Construction Company totaling NT$ 495,000,000

How has GE&L strived to become an integrated and intelligent distribution center with exceptional customer service?

By providing comprehensive logistics services and achieve complete customer satisfaction

1. _____

_____

2. _____

_____

3. _____

_____

D Based on the three situations in this unit, write three questions beginning with **What**, and answer them. The questions do not

Examples

What is Taiwan's highest health authority?

The Department of Health (DOH) of the Executive Yuan

What does DOH contribute to maintain the health of Taiwanese residents? It promulgates health-related measures to expedite the provision of convenient and efficient medical services.

1. _____

_____

2. _____

_____

3. _____

_____

Unit Four
Introducing a Company or Organization

公司或組織介紹

E Write questions that match the answers provided.

1. _____

_____

By promulgating health-related measures to expedite the provision of convenient and efficient medical services

2. _____

_____

By establishing the Bureau of Health Promotion and Bureau of Hospital Management

3. _____

_____

By establishing the National Health Education Research Group

138

F Listening Comprehension I

Situation 1

1. What did selection of Taiwan as Elsa's retail and distribution hub in the Asia Pacific region involve?

    A. an emphasis on product and service quality via a segmentation marketing strategy involving another brand

    B. careful consideration and extensive marketing research

    C. coordination of finances, factory operations and logistics

2. How long has Elsa been operating in Asia?

    A. for roughly a year

    B. for more than a year

    C. for less than a year

3. What is Sheila responsible for as an administrator in the programming division?

    A. sales and after sales service

    B. new employee orientation and on-the-job training, customer relations management-related practices and marketing strategy design

    C. implementation of marketing strategies, as well as coordination of art design projects

4. What does Elsa view as its social responsibility?

    A. the production of recyclable mattresses

    B. servicing the entire Asia Pacific region

    C. passing savings on to consumers in the form of competitive retail prices

5. How many divisions has Elsa recently established?

    A. two

    B. three

    C. four

Situation 2

1. How much did Taiwan Secom Company and Goldsun Development & Construction Company jointly invest to establish Goldsun Express & Logistics (GE&L) Company?

   A. NT$ 485,000,000

   B. NT$ 490,000,000

   C. NT$ 495,000,000

2. What is GE&L well positioned to lead?

   A. environmental resource planning

   B. the Taiwanese logistics service sector

   C. the Taiwanese integrated and intelligent distribution sector

3. How does GE&L ensure customer satisfaction through its three-stage client service approach?

   A. by tailoring offered services based on client needs

   B. by closely monitoring the flow of incoming and outgoing goods

   C. by ensuring that the logistic needs of customers are managed efficiently

4. How is the flow of incoming and outgoing goods closely monitored through the three-stage client service approach?

   A. by minimizing the influence of logistics services on business transactions during the initial service stage

   B. by outsourcing distribution services

   C. by using EIQ analysis (commonly known as volume and variation analysis)

5. Why is GE&L striving to become an integrated and intelligent distribution center with exceptional customer service?

   A. to provide comprehensive logistics services and achieve complete customer satisfaction

   B. to minimize the influence of logistics services on business transactions during the initial service stage

   C. to undertake advanced and fully automated logistics

Situation 3

1. How does DOH contribute to maintaining the health of Taiwanese residents?

    A. by encouraging diversification of hospital operations

    B. by implementing an information-based, nationwide system for preventing communicable diseases

    C. by promulgating health-related measures

2. How many outpatient services does Hsinchu General Hospital operate?

    A. 25

    B. 18

    C. 16

3. How are all individuals enabled to take increased responsibility for their well being?

    A. by establishing the National Health Education Research Group

    B. by setting up a modern health care network

    C. by promoting a healthy living space in which individuals can fully access quality healthcare services and accurate medical information

4. What is Hsinchu General Hospital's bed capacity for patients with acute chronic illnesses?

    A. 600

    B. 684

    C. 700

5. How does DOH expedite the provision of convenient and efficient medical services?

    A. by promoting the concepts of quality, efficiency and family values

    B. by offering quality medical health services to promote physical and mental well being

    C. by promulgating health-related measures

G Reading Comprehension I
Pick the work or expression whose meaning is closest to the meaning of the underlined word or expression in the following passages.

Situation 1

1. The Taiwan operations of the German mattress brand 'Elsa' were set up in 2004, with <u>aspirations</u> of servicing the entire Asia Pacific region.

   A. reproaches

   B. yearnings

   C. castigations

2. Following careful consideration and extensive marketing research, Elsa selected Taiwan as its retail and <u>distribution</u> hub in the Asia Pacific region.

   A. aggregation

   B. disbursement

   C. stockpile

3. The Taiwanese bedding market recently has experienced sales <u>growth</u>, with statistics from the Directorate General of Budget, Accounting and Statistics, Executive Yuan revealing revenue growth from NT$ 24,000,000,000 in 1991 to NT$ 59,000,000,000 in 2001.

   A. depreciation

   B. contraction

   C. augmentation

4. Besides <u>manufacturing</u> wool and cotton mattresses in Germany, in Malaysia, Elsa also produces emulsion products such as pillows and mattresses.

   A. fabricating

   B. demolishing

C. pulverizing

5. The <u>excellent</u> emulsion quality and relatively inexpensive labor in Malaysia have allowed Elsa to reduce overhead costs and pass these savings on to consumers in the form of competitive retail prices.

A. shoddy

B. mediocre

C. peerless

6. Elsa has been operating in Asia for less than a year, and has sought to <u>emphasize</u> product and service quality via a segmentation marketing strategy involving another brand.

A. refute

B. accentuate

C. discredit

7. Notably, Elsa <u>views</u> the production of recyclable mattresses as its social responsibility.

A. neglects

B. disregards

C. envisions

8. Elsa has recently <u>established</u> three divisions: accounting, business and programming.

A. instituted

B. expunged

C. ostracized

9. The accounting division coordinates finances, factory operations and logistics; the business division focuses on sales and after sales service; and the programming division devises and <u>implements</u> marketing strategies, as well as coordinates art design projects.

A. obstructs

B. promulgates

C. deters

10. As an administrator in the programming division, Sheila is responsible for new employee orientation and <u>on-the-job</u> training, customer relations management-related practices and marketing strategy design.

A. practical

B. theoretical

C. hypothetical

Situation 2

1. The result of a joint investment by Taiwan Secom Company and Goldsun Development & Construction Company totaling NT$ 495,000,000, Goldsun Express & Logistics (GE&L) Company (formerly Kuo Hsing Transport) was established in 1993 to <u>undertake</u> advanced and fully automated logistics.

A.withdraw from

B.desist from

C.commence with

2. With a licensed customs office since 2002, the company has ample skilled staff from <u>multi-disciplinary</u> fields, including security, information technology, and construction.

A. homogeneous

B. diverse

C. isolated

3. To provide comprehensive logistics services and achieve complete customer satisfaction, GE&L is striving to become an integrated and <u>intelligent</u> distribution center with exceptional customer service.

A. quarantined

B. sequestered

C. concordant

4. Well positioned to lead the Taiwanese logistics service sector, GE&L bases its success on state-of-the-art distribution procedures, information technology, automated facilities, highly skilled industrial engineering professionals, third party logistics services, full-range B2B logistics solutions and sound supply-chain management practices.

A. qualified

B. estranged

C. inhibited

5. The company deals in hi-tech products, 3C products, entertainment media (DVD, VCD, VHS), cosmetics and pharmaceuticals, all of which are products for which enterprises tend to outsource distribution services.

A. gratification

B. castigation

C. chastisement

6. GE&L has a strong organizational framework, within which logistics professionals and IT consultants jointly provide a comprehensive and diverse range of logistics services.

A. autonomously

B. concertedly

C. exclusively

7. In these joint efforts, the three-stage client service approach is adopted to ensure customer satisfaction by tailoring offered services based on client needs.

A. customizing

B. generalizing

C. universalizing

8. This approach uses EIQ analysis (commonly known as volume and variation analysis) to closely <u>monitor</u> the flow of incoming and outgoing goods.

A. glance

B. skim

C. scrutinize

9. Additionally, analyses are performed, including analysis of the industrial characteristics of goods, business transaction requirements, environmental resource planning (ERP), parameter settings and process <u>confirmation</u> for large ERP systems, and linear motion testing.

A. corroboration

B. oversight

C. dereliction

10. This procedure aims to minimize the influence of logistics services on business transactions during the <u>initial</u> service stage.

A. terminal

B. nascent

C. inevitable

11. Furthermore, the three-stage client service approach also considers how controlling inbound or outbound goods and inventory fully <u>complies with</u> the internal control system, ensuring that the logistic needs of customers are managed efficiently.

A. repudiates

B. adheres to

C. negates

Situation 3

1. As Taiwan's highest health authority, the Department of Health (DOH) of the Executive Yuan is <u>responsible for</u> administering, supervising and coordinating local health agencies.

   A. subordinate to

   B. in charge of

   C. inferior to

2. DOH contributes to <u>maintaining</u> the health of Taiwanese residents by promulgating health-related measures to expedite the provision of convenient and efficient medical services.

   A. deteriorating

   B. sustaining

   C. decaying

3. Promoting a healthy living space in which individuals can fully access quality healthcare services and accurate medical information <u>enables</u> all individuals to take increased responsibility for their well being.

   A. allows

   B. prohibits

   C. restricts

4. Key strategies implemented to achieve the above objective include establishing a National Health Insurance Review Committee to provide guidelines for <u>improving</u> the National Health Insurance system, promoting organizational reengineering by establishing the Bureau of Health Promotion and Bureau of Hospital Management, advancing national health education by establishing the National Health Education Research Group, setting up a modern health care network, enhancing health care for women and the disadvantaged, strengthening emergency medical care, encouraging diversification of hospital operations and, finally,

implementing an information-based, nationwide system for preventing communicable diseases.

A. inhibiting

B. hindering

C. enhancing

5. Key strategies implemented to achieve the above objective include establishing a National Health Insurance Review Committee to provide guidelines for improving the National Health Insurance system, promoting organizational reengineering by establishing the Bureau of Health Promotion and Bureau of Hospital Management, advancing national health education by establishing the National Health Education Research Group, <u>setting</u> <u>up</u> a modern health care network, enhancing health care for women and the disadvantaged, strengthening emergency medical care, encouraging diversification of hospital operations and, finally, implementing an information-based, nationwide system for preventing communicable diseases.

A. establishing

B. deconstructing

C. demolishing

6. Key strategies implemented to achieve the above objective include establishing a National Health Insurance Review Committee to provide guidelines for improving the National Health Insurance system, promoting organizational reengineering by establishing the Bureau of Health Promotion and Bureau of Hospital Management, advancing national health education by establishing the National Health Education Research Group, setting up a modern health care network, enhancing health care for women and the disadvantaged, strengthening emergency medical care, encouraging diversification of hospital operations and, finally, <u>implementing</u> an information-based, nationwide system for preventing

communicable diseases.

A. postponing

B. delaying

C. promulgating

7. To implement the above strategies, the Hsinchu General Hospital operates 18 outpatient services with a staff of 800, based on the concepts of quality, <u>efficiency</u> and family values.

A. efficacy

B. incompetence

C. insufficiency

8. To achieve the objectives of promulgating governmental health care policy and offering quality medical health services to promote physical and mental well being, Hsinchu General Hospital has a bed capacity of 684, with 84 intensive care beds and 600 beds for patients with acute chronic <u>illnesses</u>.

A. recovery efforts

B. remedial efforts

C. maladies

H Common elements in introducing a company or organization 公司或組織介紹 include the following:

1. Briefly overview the industry to which your organization belongs.
   簡要陳述組織所屬之產業的概況

   · Elderly in Taiwan over 65 years old accounts for 9.4% of the total population, surpassing the United Nations' definition of an aging society. Given this trend, market opportunities for senior citizen residential communities island wide are estimated at more than $US 100,000,000 annually.

   · The forecasted market demand domestically for watch purchases is 14,000 in 2005, with Rolex accounting for approximately 8.12%, or 7,714 watches with estimated revenues of nearly 120 billion New Taiwanese (NT) dollars.

   · Wanfang Hospital belongs to Taiwan's thriving medical care sector. Since its establishment in 1995, the National Health Insurance scheme has strived to provide medical coverage for all of the island's residents under the auspices of the National Health Insurance Bureau. However, by 2005, the National Health Insurance Bureau was approaching bankruptcy, making it extremely difficult to effectively manage the financial resources of hospitals.

2. Describe your organization's mission as well as highlight its historical development.
   織的使命；組織的發展沿革

   · Wanfang Hospital commits itself to the following missions: safeguard the health of Taipei City residents, strive to become the an ideal teaching hospital environment for staff and students of　Taipei Medical University and aspire to become a widely respected medical center island wide.

· Whereas older correctional facilities focused on isolating inmates from the general population without much concern for their rehabilitation, modern correctional facilities adopt a more humanistic approach towards orienting and reforming the incarcerated so that they will have practical skills upon re-entering society.

· Heavily involved in the biotechnology sector, Celsion Corporation largely focuses on advancing thermotherapy and related patented technologies to treat cancer and other diseases through the development of a thermodilatation system in order to terminate malignant tumors and treat the prostate gland.

3. Introduce the organizational structure.
   組織的架構

· With the company organized into the banking department, accounting department, information department, marketing department and general affairs department, each department strives to remain productive in order to serve customers fully.

· In addition to an administrative department, correctional facilities contain other units, such as a personnel office, accounting office, documentation and records office, civil service office, security section, rehabilitation and education section, general affairs section, health and hygiene section, general operations section, as well as many committees.

· The six bureaus that operate with National Health Insurance process insurance applications, receive insurance premiums, audit medical expenses and consult with hospitals on how to enhance their administrative management practices. Of the twenty four liaison offices located conveniently island wide to continuously upgrade the quality of medical services, 2,536 employees and 534 contract workers in 2004.

4. Highlight recent technical accomplishments within the organization.
   組織最新的科技成就

· STARE's recently developed fingerprint verification system has gained considerable recognition globally for its innovative design. Fingerprint verification systems are widely anticipated to play an increasingly prominent role in daily life given the growing incidence of computer-related crimes and theft of confidential information. Such systems could aid police in capturing perpetrators and ensuring personal confidentiality.

· While focusing on the use of heat to treat cancer and other diseases, Celsion has pioneered approaches in treating the benign swelling of the prostate gland, treating breast cancer, developing cancer treatment medicine and developing gene therapy drugs.

· Given increasing competition in the medical market sector and the severely strained financial resources of the National Health Insurance scheme, the hospital has recently expanded its services into the establishment of a Respiratory Care Center (RCC) that offers facilities for long term, chronically ill patients in order to free up bed availability in intensive care units (ICU) or other hospital resources that are in short supply.

5. Point towards the organization's future directions.

結論 （未來發展方向）

· As correctional facilities continually evolve in their approach towards equipping inmates with professional skills that will enable them to easily adjust to sociality and become productive once their prison term is completed, this more humanistic approach will hopefully ease societal problems in the long term.

· As for future directions, with governmental initiatives to liberalize and internationalize the local economy, TSC will eventually become a privatize enterprise and, in doing so, expose itself to greater competition overseas.

· While continuing with the governmental advocated policy of "Keep the root in Taiwan", the YULON group hopes that its aspirations in China and, eventually, the global automotive market will enable it to satisfy the consumer needs of the ethnic Chinese market.

In the space below, introduce a company or an organization to which your field belongs.

_____

_____

_____

_____

_____

_____

_____

_____

_____

Look at the following examples of introducing a company or an organization.

Taiwan's Ministry of Justice administers 87 correctional facilities, which take the form of prisons, training institutes, reformatory schools, detention centers, juvenile detention houses, drug abuse treatment centers and correctional high schools. However, under special conditions, juvenile correctional facilities are affiliated with the High Prosecutor's Office. Correctional facilities vary in administrative and discipline policies, depending on the target prison population — regardless of the age of the inmates. As an integral part of the justice system, correctional facilities aim not only to protect society from law offenders, but also to rehabilitate the incarcerated so that they can rejoin the general population and contribute to society. The scope of correctional facilities has dramatically changed based on legislation reform and societal values. Whereas older correctional facilities focused on isolating inmates from the general population without much concern for their rehabilitation, modern correctional facilities adopt a more humanistic approach towards orienting and reforming the incarcerated so that they will have practical skills upon re-entering society. In addition to an administrative department, correctional facilities contain other units, such as a personnel office, accounting office, documentation and records office, civil service office, security section, rehabilitation and education section, general affairs section, health and hygiene section, general operations section, as well as many committees. Juvenile correctional facilities closely resemble a normal school environment, where the incarcerated youth are allowed to develop academic and societal skills like youth in the general population. Of all the correctional facilities, minimum-security prisons are unique in that they do not heavily rely on law enforcement personnel for supervision and allow inmates to manufacture products to be sold on the market. Such facilities instill in inmates a sense of self worth and confidence, making it unlikely that they will become repeat offenders once released. As correctional facilities continually evolve in their approach towards equipping inmates with professional skills that will enable them to easily adjust to sociality and become productive once their prison term is completed, this more humanistic approach will hopefully ease societal problems in the long term.

Taiwan Sugar Corporation （TSC） has diversified its investments in an array of commodities, including sugar, processed sugar products and by-products of sugar processing, agricultural products, animal and meat products, salad oil, animal feed, beverages and even cosmetics. Established in 1946, TSC operates under the Ministry of Economic Affairs as a nationalized enterprise. Beyond the above areas, TSC has diversified into biotechnology, petroleum, merchandising, logistics, tourism, shopping malls, the construction industry and overseas investments. Moreover, as a globally

155

recognized biotechnology company, TSC designs, manufactures and develops collagen-based products, placenta and hyaluronic acid-based cosmetics products. Established as a state-run enterprise, TSC has sought to diversify its operations from its core sugar business to compete with global enterprises. Given its advances in biotechnology, TSC has applied medical-grade collagen to its biotech cosmetics, generating major revenues. As for future directions, with governmental initiatives to liberalize and internationalize the local economy, TSC will eventually become a privatize enterprise and, in doing so, expose itself to greater competition overseas. Despite these daunting challenges, TSC remains optimistic about the future. TSC perseveres in holding itself to the highest standards of excellence on behalf of its customers and the local economy. In upholding these values, TSC strives to become a globally recognized enterprise.

Originally established as YULON Machinery Co., Ltd, the enterprise's business scope focused on manufacturing machinery. Following a technical collaboration agreement with Nissan in February 1957, YULON Motor Co., Ltd. began manufacturing sedan vehicles and commercial trucks. To remain aggressive in the intensely competitive automotive sector, YULON Asia Technology Center upgraded to an engineering center in 1998, subsequently adopting a three-circle strategy to initiate the third wave of corporate restructuring. Additionally, YULON concentrated on fostering a new corporate culture based on innovation, speed and teamwork to achieve its aspiration of becoming a leader of the moving value chain in the ethnic Chinese automotive market. Responding to demands to develop a global business strategy, YULON moved its parts center from Chu-pa to Sanyi in efforts to construct the YULON Asia Parts Center in 2000 in order to become a logistics hub for China, Hong Kong and Taiwan. With its announcement to invest in the Philippines' manufacturing subsidiary of Nissan Motors Corporation in 1999, YULON entered the Southeast Asian market. Additionally, to diversify its product line, YULON formally allied itself with RENAULT in 2000, becoming its sole distributor in Taiwan. Later, in May, 2003, YULON and NISSAN jointly announced YULON's intentions to divide into two independent companies: YULON Motor Co., Ltd. and YULON-NISSAN Motor Co., Ltd. While aspiring to produce 120,000 cars in Taiwan annually and become a manufacturing service center, YULON's automotive enterprise strives to streamline customer service, increase its product line based on consumer demand and develop automobile accessories and the domestic peripherals market. Given its collaboration with NISSAN in design, research and manufacturing, YULON-NISSAN hopes to expand NISSAN operations in China's automotive market by setting a goal of selling 550,000 cars in 2006, hopefully to increase to 900,000 cars by 2009. In the future, while continuing with the governmental advocated policy of "Keep the root in Taiwan" , the YULON group hopes that its aspirations in China and, eventually, the global automotive market will enable it to satisfy the consumer needs of the ethnic Chinese market.

Situation 4

 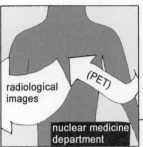 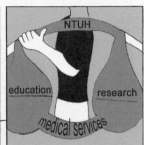

Situation 5

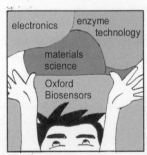 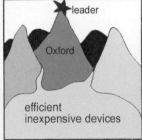 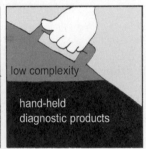

Situation 6

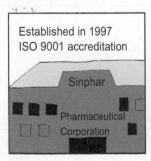  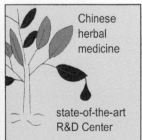

I Write down the key points of the situations on the preceding page, while the instructor reads aloud the script from the Answer Key. Alternatively, students can listen to the script online at www.chineseowl.idv. tw

Situation 4

_____

_____

Situation 5

_____

_____

_____

Situation 6

_____

_____

_____

J Oral practice II
Based on the three situations in this unit, write three questions beginning with *What*, and answer them. The questions do not need to come directly from these situations.

Examples

What does National Taiwan University Hospital (NTUH) symbolize?

The evolution of the Taiwanese medical care sector

What is NTUH committed to?

Nurturing medical talent and developing outstanding research capabilities

1. _____

_____

2. _____

_____

3. _____

_____

K Based on the three situations in this unit, write three questions beginning with *How*, and answer them. The questions do not need to come directly from these situations.

Examples

How is Oxford Biosensors Corporation committed to advancing the clinical diagnostics sector in Taiwan?

By adopting a multidisciplinary approach, employing leading research scientists in electronics, materials science, electrochemistry and enzyme technology

How has Oxford Biosensors has generated a strong and growing intellectual property portfolio?

Through successfully commercializing technologies developed in academia using funding from global investment sources

1. _____

_____

2. _____

_____

3. _____

_____

L Based on the three situations in this unit, write three questions beginning with *When*, and answer them. The questions do not need to come directly from these situations.

Examples

When was Sinphar Pharmaceutical Corporation established?

In 1977

When did Sinphar Pharmaceutical Corporation become the first pharmaceutical manufacturer in Taiwan to receive both the National Award for Small and Medium-sized Enterprises and ISO 9001 accreditation?

In 1996

1. _____

   _____

2. _____

   _____

3. _____

   _____

M Write questions that match the answers provided.

_____

_____

Both the National Award for Small and Medium-sized Enterprises and ISO 9001
accreditation

_____

_____

For both its factory operations and product innovation

_____

_____

Over NT$ 230,000,000

N Listening Comprehension II

Situation 4

1. How long has National Taiwan University Hospital (NTUH) been operating?

   A. nearly a century

   B. more than a century of history

   C. more than a century and a half

2. What is NTUH globally renowned as?

   A. a premier teaching hospital in Asia

   B. a premier nuclear medicine center in Asia

   C. a premier cancer therapy center in Asia

3. What is NTUH committed to nurturing?

   A. nursing home care services

   B. a strong medical curriculum

   C. medical talent

4. How does the nuclear medicine department provide radiological images for diagnosis?

   A. by producing general diagnostic images using x-ray MRI examinations

   B. by injecting radiological pharmaceuticals and using advanced methods such as PET

   C. by providing therapies such as surgery, chemotherapy and radiation therapy

5. Why does NTUH actively encourage international cooperation?

   A. to set precedents for the entire medical community

   B. to yield optimal research results

   C. to remain abreast of global trends in medicine

Situation 5

1. In what areas does Oxford Biosensors Corporation employ leading research scientists

   A. healthcare

   B. electronics, materials science, electrochemistry and enzyme technology

   C. biosensor development

2. What is an example of a technical breakthrough pioneered at Oxford Biosensors?

   A. medical diagnostic laboratory instrumentation

   B. a diverse range of low complexity, hand-held diagnostic products

   C. bioanalytical procedures

3. How has Oxford Biosensors generated a strong and growing intellectual property portfolio?

   A. by closely collaborating on results-oriented endeavors

   B. by drawing heavily upon the personnel and resources of Oxford University

   C. through successfully commercializing technologies developed in academia using funding from global investment sources

4. When was the pilot plant manufacturing facility of Biosensors established?

   A. in 2000

   B. in 2001

   C. in 2002

5. Why is the electrochemistry program at Oxford Biosensors particularly noteworthy?

   A. Significant resources have been invested in commercializing biosensor technologies.

   B. Oxford Biosensors provides medical diagnostic laboratories with accurate, efficient and inexpensive devices.

   C. Internationally recognized experts comprise the corporate advisory board.

Situation 6

1. What did Sinphar Pharmaceutical Corporation receive in 1996?

    A. ISO 17025 accreditation

    B. GLP accreditation

    C. ISO 9001 accreditation

2. How much did Sinphar invest Sinphar to construct its research and development center in Taiwan?

    A. nearly NT$ 230,000,000

    B. over NT$ 230,000,000

    C. approximately NT$ 230,000,000

3. What demonstrates the commitment of the Sinphar Group to developing state-of-the-art product technologies?

    A. a 12,000 square feet, five-story facility

    B. An R&D budget of $NT 80,000,000 for 2003

    C. its clinical trial facility for Chinese herbal medicine

4. What greatly facilitates Sinphar's product technology research?

    A. a state-of-the-art R&D Center

    B. further investment of up to $NT 450,000,000

    C. its biotechnology expertise to extract and purify traditional Chinese medicine components

5. When was Sinphar Pharmaceutical Corporation established?

    A. in 2002

    B. in 2003

    C. in 1977

O Reading Comprehension II
Pick the work or expression whose meaning is closest to the meaning of the underlined word or expression in the following passages.

Situation 4

1. With more than a century of history, National Taiwan University Hospital (NTUH) symbolizes the <u>evolution</u> of the Taiwanese medical care sector.

   A. retrogression

   B. fruition

   C. reversion

2. The hospital has two main buildings, and the western wing was the largest and most modern hospital in Southeast Asia upon its <u>construction</u> in 1898.

   A. erection

   B. demolition

   C. eradication

3. Committed to nurturing medical talent and developing outstanding research capabilities, NTUH strives to set <u>precedents</u> for the entire medical community.

   A. repetition

   B. recurrences

   C. paradigms

4. Given the <u>inestimable</u> value of life and the growth in global health consciousness, NTUH is globally renowned as a premier teaching hospital in Asia.

   A. sure-thing

   B. predictable

   C. fathomless

5. Regarding the <u>organizational</u> structure of NTUH, decision making is performed by the secretarial department.

A. systematic

B. unruly

C. disjoined

6. Meanwhile, the administrative department <u>comprises</u> the financial, personnel, accounting, information technology and other smaller units.

A. lacks

B. embodies

C. falls short of

7. Furthermore, the medical departments include the nutritional, pharmaceutical, nursing and <u>aesthetic</u> units, as well as the hepatitis research center.

A. materialistic

B. artistic

C. sedentary

8. The medical departments also include various units within the hospital, including three major departments <u>involved</u> in medical radiology.

A. refrained

B. abstained

C. engaged

9. First, the nuclear medicine department provides radiological images for diagnosis by <u>injecting</u> radiological pharmaceuticals and using advanced methods such as PET.

A. inserting

B. extracting

C. extricating

10. Second, the medical imaging department produces general diagnostic images using x-ray MRI <u>examinations</u>.

    A. oversights

    B. dereliction

    C. investigations

11. Third, the cancer <u>therapy</u> center provides therapies such as surgery, chemotherapy and radiation therapy.

    A. treatment

    B. prevention

    C. aversion

12. NTUH is recognized as an international <u>leader</u> in several fields, especially hepatitis, organ transplantation, nasal and paralegals cancer, liver and stomach cancer therapy and antivenin research.

    A. innovator

    B. apprentice

    C. emulator

13. As a teaching hospital, NTUH stresses education and research as well as medical services. In terms of education, the hospital seeks to <u>foster</u> medical talent in various fields by following a strong medical curriculum.

    A. constrain

    B. nurture

    C. inhibit

14. As for research, NTUH <u>integrates</u> the resources of various fields and adopts state-of-the-art equipment to yield optimal research results.

    A. segment

    B. partition

    C. amalgamate

15. Regarding medical services, the hospital focuses on serving patient needs and <u>strengthening</u> its organizational structure.

    A. taming

    B. invigorating

    C. domesticating

16. Besides offering nursing home care services, NTUH actively encourages international cooperation to remain <u>abreast of</u> global trends in medicine.

    A. in tune with

    B. opposed to

    C. distrustful of

Situation 5

1. <u>Committed</u> to advancing the clinical diagnostics sector in Taiwan, Oxford Biosensors Corporation adopts a multidisciplinary approach, employing leading research scientists in electronics, materials science, electrochemistry and enzyme technology.

    A. reluctant

    B. obliged

    C. hesitant

2. Internationally recognized experts comprise the corporate advisory board, and the corporation draws heavily upon the personnel and resources of Oxford University, which is <u>renowned</u> worldwide for its innovations in electrochemistry and biosensor development.

    A. enigmatic

    B. illustrious

    C. concealed

3. Through successfully commercializing technologies developed in academia using underline{funding} from global investment sources, Oxford Biosensors has generated a strong and growing intellectual property portfolio, based on a commitment to product development and broad expertise in biosensor design, development and manufacturing.

A. capital

B. facilities

C. equipment

4. Specifically, Oxford Biosensors provides medical diagnostic laboratories with underline{accurate}, efficient and inexpensive devices, and is a leader in this field.

A. amiss

B. erroneous

C. precise

5. Effective healthcare requires accurate data regarding certain biochemical parameters, meaning that biosensors play a crucial role in modern healthcare owing to their specificity, underline{miniaturized} size, rapid response and relative inexpensiveness.

A. diminutive

B. magnified

C. amplified

6. Bioanalytical procedures are widely underline{anticipated} to be increasingly adopted for measuring metabolites, blood cations and gases.

A. forecasted

B. nullified

C. annulled

7. Established in 2000 based on technological innovations by Oxford University, the underline{pilot} plant manufacturing facility of Biosensors is located in nearby Yarnton,

England.

A. perpetual

B. incessant

C. trial

8. As a rapidly growing healthcare diagnostics development company, Oxford Biosensors provides a challenging and <u>stimulating</u> work environment for entrepreneurial individuals committed to constant technological innovation.

A. tedious

B. monotonous

C. invigorating

9. Company employees have a strong sense of how they can contribute by closely collaborating on results-oriented <u>endeavors</u>.

A. vacancies

B. undertakings

C. voids

10. Oxford Biosensors <u>focuses</u> mainly on electrochemistry and enzyme technology.

A. diverts

B. zeros in on

C. shunts

11. The electrochemistry program is particularly <u>noteworthy</u> as significant resources have been invested in commercializing biosensor technologies.

A. exceptional

B. irrelevant

C. trivial

12. The technical <u>breakthroughs</u> pioneered at Oxford Biosensors include the development of a diverse range of low complexity, hand-held diagnostic products.

A. stumbling blocks

B. bottlenecks

C. innovations

13. For example, the Multisense Dry Enzyme System satisfies consumer demand for rapid and accurate multi-parameter analysis by instantly providing <u>essential</u> diagnostic results.

A. requisite

B. obsolete

C. oblivious

Situation 6

1. Established in 1977, Sinphar Pharmaceutical Corporation became the first pharmaceutical manufacturer in Taiwan to receive both the National Award for Small and Medium-sized Enterprises and ISO 9001 <u>accreditation</u> in 1996.

A. interdiction

B. validation

C. rebuff

2. Later, in 2002, Sinphar was the only pharmaceutical manufacturer to receive the National Biotechnology & Medical Care <u>Award</u> for both its factory operations and product innovation.

A. reprimand

B. recognition

C. reproof

3. After <u>investing</u> over NT$ 230,000,000, Sinphar constructed the first research and development center in Taiwan to receive ISO 17025 and GLP accreditation.

A. extracting

B. withdrawing

C. patronizing

4. Licensed for operations in July of 2002, with production beginning in October of that year, the 12,000 square feet, five-story facility became a model for <u>integrating</u> domestic and international biotechnology research.

A. consolidating

B. discriminating between

C. uncoupling

5. An R&D budget of $NT 80,000,000 for 2003, along with further investment of up to $NT 450,000,000 over the next five years, demonstrates the <u>commitment</u> of the Sinphar Group to developing state-of-the-art product technologies.

A. revocation

B. covenant

C. nullification

6. An R&D <u>budget</u> of $NT 80,000,000 for 2003, along with further investment of up to $NT 450,000,000 over the next five years, demonstrates the commitment of the Sinphar Group to developing state-of-the-art product technologies.

A. arrears

B. shortfall

C. operating expenses

7. Committed to continuously upgrading business operations, <u>expanding</u> upon patented technological innovations, and ensuring superior product quality, Sinphar utilizes its biotechnology expertise to extract and purify traditional Chinese medicine components in order to create new products for clinical trials and eventual commercialization.

A. escalating

B. constricting

C. diminishing

8. Committed to continuously upgrading business operations, expanding upon patented technological innovations and ensuring <u>superior</u> product quality, Sinphar utilizes its biotechnology expertise to extract and purify traditional Chinese medicine components in order to create new products for clinical trials and eventual commercialization.

A. menial

B. nonparallel

C. second-fiddle

9. A state-of-the-art R&D Center greatly facilitates product technology research. A notable example is its clinical trial facility for Chinese herbal medicine, which applies advanced scientific approaches to extract and <u>purify</u> components of Chinese herbal medicine and, eventually, to standardize formulations for clinical studies.

A. mar

B. sanitize

C. deprave

10. A state-of-the-art R&D Center greatly facilitates product technology research. A notable example is its clinical trial facility for Chinese herbal medicine, which applies advanced scientific approaches to <u>extract</u> and purify components of Chinese herbal medicine and, eventually, to standardize formulations for clinical studies.

A. extirpate

B. infuse

C. inoculate

Unit Five

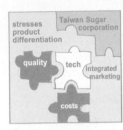
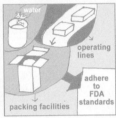
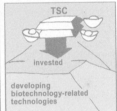

Introducing a Division or a Department

組或部門介紹

1. Describe the setting of your
   division or department within the largerorganization.
   介紹部門所屬的組織或公司

2. Highlight the organizational structure of the division or department.
   部門的組織架構

3. Point out the staff's strengths and educational backgrounds/ work experience in a
   particular field.
   部門的人才來源和教育背景

4. Spell out the missions of the division or department.
   部門的使命

5. Elaborate on the manufacturing or research capabilities within the division or
   department.
   部門之製造或研究的能力

6. List the technical services (e.g. industrial and consultancy) that the department or
   division offers.
   部門提供的產業服務

Vocabulary and related expressions

integrates the efforts of	整合努力的成果
continuously upgrade	持續的提升
intensely competitive	強烈地競爭
constantly changing market	不斷變化的市場
fulfill their career aspirations	滿足其職業抱負
in line with corporate goals	跟公司全體的目標一致
widely anticipated	廣泛地預期
high value-added products	有高度附加價值的商品
core competence	核心價值
commercial viability	商業的可行性
intensely competitive	高度競爭
logistics sector	物流部門
a cohesive business strategy	凝聚性的商業策略
accumulated extensive experience	累積的大規模的經驗
closely collaborates with	密集和……合作
independent problem solving skills	單獨解決問題的技能
numerous management opportunities	多個的管理機會
gain consumer confidence	獲得消費者的信任
adopting the latest technological practices	採取最新的科技運用
commitment to excellence	承諾會表現傑出
highly talented staff	具有高度天分的員工
a nurturing environment	有滋養的環境
coping with	和……合作
specializing in	使專門化
under the auspices of	受……的贊助
As for future trends	至於未來的趨勢

Situation 1

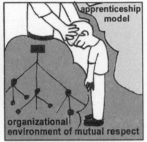

Situation 2

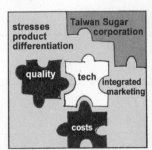
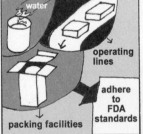
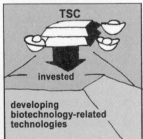

Situation 3

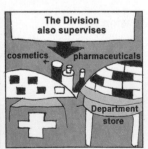

A Write down the key points of the situations on the preceding page, while the instructor reads aloud the script from the Answer Key. Alternatively, students can listen to the script online at www.chineseowl.idv. tw

Situation 1

_____

_____

Situation 2

_____

_____

_____

Situation 3

_____

_____

_____

B Oral practice I

Based on the three situations in this unit, write three questions beginning with *How*, and answer them. The questions do not need to come directly from these situations.

Examples

How does the Educational Training and Orientation Department at China Motors Corporation upgrade the professional skills of company employees?

Through a comprehensive education and training curricula

How does the Department assess the talents of employees and recommend ways of better utilizing their professional skills and potential?

By closely following the missions and planning strategy of CMC

1. _____

   _____

2. _____

   _____

3. _____

   _____

C Based on the three situations in this unit, write three questions beginning with **Why**, and answer them. The questions do not need to come directly from these situations

Examples

Why has the Sales Department at Taiwan Sugar Corporation (TSC) adopted an integrated marketing approach that stresses product differentiation?

To stay in line with corporate goals of enhancing biotechnology product quality, reducing overhead costs and adopting the latest professional technologies

Why is the Biotechnology Division preparing to install extraction equipment for water and organic solvents, operating lines and packing facilities?

To manufacture various high value-added products that adhere to FDA standards

1. _____

_____

2. _____

_____

3. _____

_____

D Based on the three situations in this unit, write three questions beginning with **What**, and answer them. The questions do not need to come directly from these situations.

Examples

What is the smallest division in Goldsun Express & Logistics Corporation in terms of employees?

The Service Division

What does the Service Division strive to do?

Give Goldsun Express & Logistics Corporation a competitive edge in Taiwan's intensely competitive logistics sector

1. _____

_____

2. _____

_____

3. _____

_____

E Write questions that match the answers provided.

1. _____

   _____

   For a decade

2. _____

   _____

   Five

3. _____

   _____

   An average of two years

F Listening Comprehension I

Situation 1

1. How does the Educational Training and Orientation Department continuously upgrade the professional skills of company employees?

   A. through closely following the missions and planning strategy of China Motors Corporation

   B. through implementing an apprenticeship model in which new employees can work in specific areas under the guidance of more seasoned colleagues

   C. through a comprehensive education and training curricula

2. How does the Department fully orient new employees regarding company operations and equip them with the professional skills necessary to perform daily tasks?

   A. by maintaining an employee training center

   B. by maintaining a consultation room that allows colleagues to freely discuss professional and personal concerns

   C. by encouraging employees to continuously upgrade their reading skills to maintain their professional skills

3. Why does the employee training center often consult with its counterparts in industry and government?

   A. regarding how to foster an organizational environment of mutual respect

   B. regarding how to maintain the professional skills of employees

   C. regarding how to improve its operations

4. Why does the Educational Training and Orientation Department assess the talents of employees and recommends ways of better utilizing their professional skills and potential?

   A. to better fulfill the career aspirations of employees while maintaining a healthy family and personal life

B. to more effectively respond to the intensely competitive and constantly changing market

C. to encourage employees to maintain their professional skills

5. What does the Educational Training and Orientation Department integrate?

A. the efforts of human resources personnel, training instructors and library employees

B. the professional skills of company employees through a comprehensive education and training curricula

C. specific activities to continuously upgrade the professional skills of company employees

Situation 2

1. Why is the Biotechnology Division preparing to install extraction equipment for water and organic solvents, operating lines and packing facilities?

A. to increase its technical expertise in fermentation, extraction, biological reagents, biomedical materials, biopharmaceuticals and in vitro diagnostics

B. to remain flexible and highly responsive to market fluctuations

C. to manufacture various high value-added products that adhere to FDA standards

2. In what area is the Sales Department committed to educating customers?

A. regarding the health benefits of its biotechnology products

B. regarding its core competence and commercial viability in functional foods and cosmetics

C. regarding its expertise in fermentation, extraction, biological reagents, biomedical materials, biopharmaceuticals and in vitro diagnostics

3. How has the Sales Department at Taiwan Sugar Corporation stayed in line with corporate goals of enhancing biotechnology product quality, reducing overhead costs and adopting the latest professional technologies?

   A. by developing product technologies and technical expertise in fermentation, extraction, biological reagents, biomedical materials, biopharmaceuticals and in vitro diagnostics

   B. by adopting an integrated marketing approach that stresses product differentiation

   C. by adhering to FDA standards

4. When are advanced biotechnology products widely anticipated to reach commercialization?

   A. within the next year

   B. within the next two years

   C. within the next five years

5. Why has Taiwan Sugar Corporation invested heavily in developing biotechnology-related technologies?

   A. to meet expected strong future market demand

   B. to determine its core competence and commercial viability in the biotechnology market

   C. to ensure corporate success

Situation 3

1. How large is the Service Division of Goldsun Express & Logistics Corporation?

    A. the second to largest division in the company in terms of employees

    B. the smallest division in the company in terms of employees

    C. the largest division in the company in terms of employees

2. What is the average professional experience in the logistics sector for employees in the Service Division?

    A. more than two years

    B. nearly two years

    C. two years

3. What is the company's business strategy based on?

    A. customer relations management-related practices

    B. maintaining a competitive edge in Taiwan's intensely competitive logistics sector

    C. devising and implementing marketing strategies, as well as coordinating customer design projects

4. What products does the Service Division supervise delivery to regional hospitals, medical centers and department stores?

    A. 3C products such as mobile phones, personal computers (PC) and PC peripherals, personal digital assistants (PDA)

    B. PDA peripherals and entertainment media products, including DVDs, VCDs, CDs and VHS tapes

    C. cosmetics and pharmaceutical products

5. How long has the Service Division of Goldsun Express & Logistics Corporation been in existence?

    A. for more than a decade

    B. for two decades

    C. for a decade

G Reading Comprehension I
Pick the work or expression whose meaning is closest to the meaning of the underlined word or expression in the following passages.

Situation 1

1. The Educational <u>Training</u> and Orientation Department at China Motors Corporation (CMC) integrates the efforts of human resources personnel, training instructors and library employees.

   A. interrogating

   B. debriefing

   C. orienting

2. The Department performs specific activities to continuously upgrade the professional skills of company employees through a <u>comprehensive</u> education and training curricula.

   A. capacious

   B. primordial

   C. rudimentary

3. First, while closely following the missions and planning strategy of CMC, the Department assesses the talents of employees and recommends ways of better utilizing their professional skills and potential to more effectively respond to the intensely competitive and constantly <u>changing</u> market.

   A. immutable

   B. vacillating

   C. abiding

4. First, while closely following the missions and planning strategy of CMC, the Department assesses the <u>talents</u> of employees and recommends ways of better

utilizing their professional skills and potential to more effectively respond to the intensely competitive and constantly changing market.

A. flaws

B. imperfections

C. ingenuity

5. Second, the Department maintains an employee training center to fully <u>orient</u> new employees regarding company operations and equip them with the professional skills necessary to perform daily tasks.

A. indoctrinate

B. delude

C. hoodwink

6. The center often <u>consults with</u> its counterparts in industry and government regarding how to improve its operations.

A. confers with

B. locks horns with

C. quibbles with

7. Third, the Department maintains a consultation room that allows colleagues to <u>freely discuss</u> professional and personal concerns regarding how to better fulfill their career aspirations while maintaining a healthy family and personal life.

A. squelch

B. effectuate

C. quell

8. Third, the Department maintains a consultation room that allows colleagues to freely discuss professional and personal concerns regarding how to better fulfill their career aspirations while <u>maintaining</u> a healthy family and personal life.

A. curtailing

B. prolonging

C. retrenching

9. Finally, the Department maintains a well stocked library containing numerous books, periodicals and multimedia teaching materials, thus encouraging employees to <u>continuously</u> upgrade their reading skills to maintain their professional skills.

A. intermittently

B. spasmodically

C. perennially

10. Moreover, the Department implements an apprenticeship model in which new employees can work in specific areas under the guidance of more <u>seasoned</u> colleagues, thus fostering an organizational environment of mutual respect.

A. callow

B. acclimated

C. verdant

11. Moreover, the Department implements an apprenticeship model in which new employees can work in specific areas under the guidance of more seasoned colleagues, thus fostering an organizational environment of <u>mutual</u> respect.

A. analogous

B. discordant

C. antagonistic

Situation 2

1. In line with corporate goals of enhancing biotechnology product quality, reducing overhead costs and adopting the latest professional technologies, the Sales Department at Taiwan Sugar Corporation (TSC) has adopted an <u>integrated</u> marketing approach that stresses product differentiation.

   A. incompetent

   B. pragmatic

   C. rudimentary

2. In line with corporate goals of enhancing biotechnology product quality, reducing overhead costs and adopting the latest professional technologies, the Sales Department at Taiwan Sugar Corporation (TSC) has adopted an integrated marketing approach that stresses product <u>differentiation</u>.

   A. similitude

   B. distinction

   C. parallel

3. Advanced biotechnology products are widely <u>anticipated</u> to reach commercialization within the next two years.

   A. predicted

   B. misconstrued

   C. misinterpreted

4. The Biotechnology Division of TSC focuses on developing product technologies and technical <u>expertise</u> in fermentation, extraction, biological reagents, biomedical materials, biopharmaceuticals and in vitro diagnostics.

   A. callowness

   B. greenness

   C. adeptness

5. The Division is preparing to <u>install</u> extraction equipment for water and organic

solvents, operating lines and packing facilities to manufacture various high value-added products that adhere to FDA standards.

A. pluck out

B. eradicate

C. embed

6. Since research and development, procurement, production, sales and technical services are all <u>integral</u> to corporate success, the organizational structure of TSC is both flexible and highly responsive to market fluctuations.

A. intrinsic

B. expendable

C. dispensable

7. Given the recent <u>emergence</u> of biotechnology, TSC has invested heavily in developing biotechnology-related technologies to meet expected strong future market demand, as demonstrated by the Taiwanese government policy of encouraging the local private sector to enter this field.

A. revocation

B. secession

C. efflux

8. Given the recent emergence of biotechnology, TSC has invested heavily in developing biotechnology-related technologies to meet expected <u>strong</u> future market demand, as demonstrated by the Taiwanese government policy of encouraging the local private sector to enter this field.

A. negligibly

B. prodigiously

C. scarcely

9. The Sales Department adopts flexible procedures for determining the core <u>competence</u> and commercial viability of TSC in the biotechnology market,

especially in functional foods and cosmetics.

A. ineptitude

B. inefficacy

C. aptitude

10. Moreover, the Department is committed to educating customers regarding the health <u>benefits</u> of TSC's biotechnology products.

A. merits

B. liabilities

C. impediments

Situation 3

1. In existence for a decade, the Service Division of Goldsun Express & Logistics Corporation (formerly Kuo Hsing Transport) comprises ten employees <u>dedicated</u> to devising and implementing marketing strategies, as well as coordinating customer design projects.

A. unresponsive

B. indifferent

C. devoted

2. The smallest division in the company in terms of employees, the Service Division also operates a service center, <u>staffed</u> by five service engineers and an engineering assistant, all with undergraduate degrees related to their areas of experience and an average of two years professional experience in the logistics sector.

A. employed

B. vacated

C. abandoned

3. The Service Division <u>strives</u> to give Goldsun Express & Logistics Corporation a competitive edge in Taiwan's intensely competitive logistics sector by focusing

on customer service, quality management, service quality and customer satisfaction.

A. dawdles

B. lags

C. aspires

4. The Service Division strives to give Goldsun Express & Logistics Corporation a competitive edge in Taiwan's <u>intensely</u> competitive logistics sector by focusing on customer service, quality management, service quality and customer satisfaction.

A. tediously

B. monotonously

C. fervently

5. Although not directly generating revenue for the company, the Division is still indispensable in daily operations, integrating the efforts of other divisions to execute a <u>cohesive</u> business strategy based on customer relations management-related practices.

A. congruent

B. sketchy

C. truncated

6. Although not directly generating revenue for the company, the Division is still <u>indispensable</u> in daily operations, integrating the efforts of other divisions to execute a cohesive business strategy based on customer relations management-related practices.

A. optional

B. requisite

C. elective

7. For instance, logistics services require <u>punctual</u> product delivery to satisfy customers. In terms of daily operations, the Service Division handles electronic-based orders, current accounts, delivery, stock inventory control, stock inventory-related data and delivery of hi tech products to wafer fabs and retail shops.

A. lingering

B. immediate

C. procrastinated

8. For instance, logistics services require punctual product delivery to <u>satisfy</u> customers. In terms of daily operations, the Service Division handles electronic-based orders, current accounts, delivery, stock inventory control, stock inventory-related data and delivery of hi tech products to wafer fabs and retail shops.

A. appease

B. annoy

C. hector

9. Products include 3C products such as <u>mobile</u> phones, personal computers (PC), PC peripherals, personal digital assistants (PDA), PDA peripherals and entertainment media products, including DVDs, VCDs, CDs and VHS tapes.

A. motionless

B. stationary

C. portable

10. The Division also <u>supervises</u> delivery of cosmetics and pharmaceutical products to regional hospitals, medical centers and department stores.

A. orchestrates

B. shuns

C. disregards

H Common elements in introducing a division or a department / 組或部門介紹 include the following contents:

1. Describe the setting of your division or department within the larger organization.

   介紹部門所屬的組織或公司

· Although not directly generating revenues for the company, the Programming Division still plays an indispensable in daily operations by integrating the efforts of other divisions within Elsa to execute a cohesive business strategy based on customer relations management-related practices (CRM) and other marketing strategies.

· The Bureau of Monetary Affairs (BMA) of the Financial Supervisory Commission operates under the Executive Yuan of the Republic China. Established for a century and recently renamed in 2004, the BMA is a governmental department whose employees must pass a nationwide civil service examination to gain employment, which is extremely competitive.

· Within the company, the Enterprise Planning Department, Recycling Department, Environmental Engineering Department and Engineering Maintenance Department handle daily operations, as staffed by ninety employees. Responsible for handling waste efficiently and safely, the Recycling Department treats infectious waste, waste from the steel industry and heavy metals.

2. Highlight the organizational structure of the division or department.
部門的組織架構

· The regional bureau consists of seven sections and four offices: first underwriting section, second underwriting section, third underwriting section, medical affairs section, outpatient expenditure section, inpatient expenditure section, medical expenditure review, personnel office, information management office, civil service ethics office and accounting office - all of which are under the direction of the General Secretariat.

· The department consists of the following six working groups: biochemical, blood work, serum, blood bank, blood examination and bacterium. Departmental work space is divided into three areas: the hospital's comprehensive examination room, outpatient services and an emergency unit.

· As a leading cord blood bank in Taiwan with the safest cord blood cryopreservation facility in Asia, Sino Cell Technologies (SCT) provides high quality cord blood stem cell processing and storage capabilities. Within the organization, the Cord Blood Registry processes, tests and stores cord blood stem cells, assuming responsibility for processing, quality control and quality-assurance metrics that comply with FDA guidelines.

3. Point out the staff's strengths and educational backgrounds/ work experience in a particular field.
部門的人才來源和教育背景

· As for the personnel, programming and coding designers, artists, music and game designers, as well as quality controllers comprise 50%, 20%, 10% and 10%, respectively, form the crux of the staff whom have an average working experience of five years. The broad range of educational levels, from programming certifications to doctorate degrees, enables our department, Programming and Coding Design, to handle complicated design tasks.

· Comprising a departmental head, four medical radiology technologists five professional nurses and an administrative assistant, CR staff have at least a junior college diploma from a medical college or higher. Recruited through the human resources division via newspaper advertisements and a rigorous interview process, departmental staff adopts a professional and congenial attitude towards serving patients as family members, regardless of the severity of their condition.

· Headed by a director, the department comprises two senior engineers and three operators, with each acquiring at least a bachelor's degree and an average of five years professional experience.

4. Spell out the missions of the division or department.
部門的使命

· While devoting its energies to devising and implementing marketing strategies, as well as coordinating art design projects, the Programming Division strives to give Elsa a competitive edge in the bedding sector by adopting sound management practices that will ensure customer satisfaction.

· Our subsection is largely responsible for coordinating customer services, including the issuing of IC cards, processing insurance for newborns, calculating and adjusting insurance premiums, processing group insurance applications, supervising the activities of a heart donor fund for the economically disadvantaged and overseeing installment payments.

· Specifically, the department strives to generate company revenues, promote the entertainment quality of company products, instill in customers a sense of concern and respect for their particular viewing or listening tastes and foster a harmonious home environment with the latest DVD or VCD movie releases.

5. Elaborate on the manufacturing or research capabilities within the division or department.
部門之製造或研究的能力

· The Department adopts advanced technological equipment imported from Japan that melts infectious waste at high temperatures. In this process, infectious waste is collected and then treated at combustion temperatures ranging from 1650°C~3000°C. Using this procedure, the Recycling Department services hospitals, schools, airports, hotels and community disposal facilities.

· While striving to develop on-line games for commercialization, the department heavily relies on the highly qualified staff that, in addition to designing games for consumer use, has developed a 3D software program to create realistic images.

· Areas of particular interest that the department is involved includes the dosage rate at which cancer cells can totally resist the effects of a therapeutic drug, the rate at which normal cells are terminated during drug dosage, the ability to achieve a survival rate of 50% for cells during the drug dosage, ability to protect normal cells from chemotherapy drugs, a more thorough understanding of how cancer cells are resistant to chemotherapy drugs and a criterion for medicine dosage.

6. List the technical services (e.g. industrial and consultancy) that the department or division offers.
部門提供的產業服務

· The Healthcare Center takes full advantage of its numerous years of solid experience by focusing on the health concerns of the elderly, implementing health sustaining programs, providing individual counseling on how to effectively address the physiological and psychological needs of the elderly and, ultimately, establish a sustainable healthcare system that caters to the special needs of Taiwan's aging population.

· The Bureau of Monetary Affairs is divided into six units to handle different aspects of daily operations: monitoring the development of financial organizations, making relevant regulations, supervising financial market reforms, addressing stock-related problems, analyzing financial events and facilitating mergers of financial organizations.

· Given its pivotal role in the hospital's daily operations, the Radiology Department runs a general radiology area with portable instrumentation and four x-ray examination rooms that take a) routine x-ray images of specific areas such as the chest, abdomen, skull, limbs and spine. b) unique x-ray images such as the lower and upper GI, c) computed tomography images for head, abdominal and spinal scans which are common for the emergency care department and the orthopedics department and d) ultrasound images for patients requiring sonar scans of the hand and abdomen.

In the space below, introduce a division or department of a company

_____

_____

_____

_____

_____

_____

_____

_____

_____

Look at the following examples of introducing a division or a department in a company.

Sedentary lifestyles, hectic work schedules and lack of athletic activity contribute to poor physical postures, shoulder aches and back pain. Several Japanese physicians attribute 80% of all chronic illnesses to vertebrae-related problems, which can be remedied by physical activity and sufficient rest. As a manufacturer of physical exercise equipment, the Taiwanese branch of OSIM began exporting its products worldwide in 2004. More than just offering massage equipment, OSIM has several patents for its product technologies, which come in a diverse array of product types and colors. With 66 retail outlets island wide, OSIM has generated record profits for the burgeoning consumer market. In addition to its patented technologies worldwide, OSIM stresses the importance of after-sales service for all of its products to retain consumer confidence. Additionally, its recently established research and development center focuses on product innovation to offer diversity in its product offerings and gain a market niche among increasingly health conscious consumers. As potential customers range from the young to the elderly, health promotion is the emphasis of all marketing strategies. Given the growing health consciousness globally, OSIM has recently expanded outside of Southeast Asia into countries such as Australia, South Africa and the United States.

Established for less than a year, the Programming Division of the Elsa Bedding Franchise in Taiwan consists of five employees whom devote their energies to devising and implementing marketing strategies, as well as coordinating art design projects. In contrast to eight other colleagues involved in manufacturing, ten in logistics, twenty in sales and eight in accounting, this division is the smallest in terms of employees. As an administrator in this division, I work with two other administrators and two administrative assistants, all having an average of three years of professional experience in the furniture and retail sector, as well as educational training at the undergraduate and graduate levels. The Programming Division strives to give Elsa a competitive edge in the bedding sector by adopting sound management practices that will ensure customer satisfaction. Given the relatively recent emergence of the bedding sector in Taiwan, Elsa focuses on creating its own unique brand recognition. Although not directly generating revenues for the company, the Programming Division still plays an indispensable role in daily operations by integrating the efforts of other divisions within Elsa to execute a cohesive business strategy based on customer relations management-related practices (CRM) and other marketing strategies. For instance, CRM entails the Sales Division to accumulate purchasing data of customers through data mining and other means, the Accounting Division to maintain a customer data base, the Manufacturing Division to produce high quality products based on consumer preferences and the Logistics Division to ensure timely delivery of products. CRM would obviously fail if each division did not to integrate its efforts with those of the other divisions. In another development, the Programming Division launched the commercial promotion of the electronic bed 2036 series, in which the company's logistics process must conform to standard operating procedures during manufacturing and each bedding outlet in Taiwan must maintain a management-oriented customer database. So far, the company has reaped approximately $NT 3,000,000 in profits, enhanced customer satisfaction by 50% owing to the enhanced operating procedures, dramatically reduced office overhead costs via computerization and increased the efficiency of data transmission.

As a compulsory social insurance program for all Taiwanese residents, the National Health Insurance (NHI) system provides public access to advanced and equitable medical care. While striving to maintain as comprehensive medical coverage as possible, the NHI Bureau closely operates under the auspices of the governing body Executive Yuan to maintain a fiscal balance. The NHI program aggressively strives to achieve the following: universal medical coverage for all of Taiwan's residents, equitable and proportional medical care island wide, fiscal responsibility to avoid a deficit incurred from medical treatment, long term operations and quality medical care. The Northern Region Bureau of the NHI to which I belong is responsible for servicing Taoyuan, Hsinchu and Miaoli residents, or roughly 3,000,000 inhabitants. The regional bureau consists of seven sections and four offices: first underwriting section, second underwriting section, third underwriting section, medical affairs section, outpatient expenditure section, inpatient expenditure section, medical expenditure review, personnel office, information management office, civil service ethics office and accounting office - all of which are under the direction of the General Secretariat. Specifically, I belong to the second of three subsections of the first underwriting section. Our subsection is largely responsible for coordinating customer services, including the issuing of IC cards, processing insurance for newborns, calculating and adjusting insurance premiums, processing group insurance applications, supervising the activities of a heart donor fund for the economically disadvantaged and overseeing installment payments. To enhance service quality of NIH and broaden its scope of community services, our subsection also trains retired individuals to contribute to the organization. In sum, our subsection continuously strives to enhance the NHI system and fully serve our clients' needs, enabling NHI to operate in a sustainable manner and elevate the living standards of Taiwanese residents.

Situation 4

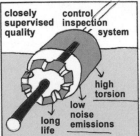

Situation 5

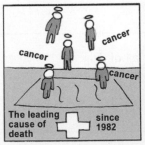

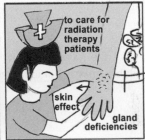

Situation 6

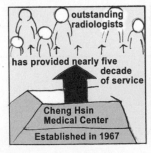

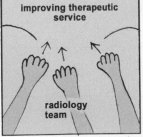

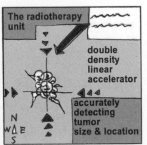

203

I Write down the key points of the situations on the preceding page, while the instructor reads aloud the script from the Answer Key. Alternatively, students can listen to the script online at www.chineseowl.idv. tw

Situation 4

_____

_____

_____

Situation 5

_____

_____

_____

Situation 6

_____

_____

_____

J Oral practice II

Based on the three situations in this unit, write three questions beginning with *What*, and answer them. The questions do not need to come directly from these situations.

Examples

What has Yuan Electrical Machinery Company accumulated extensive experience in? Developing high-accuracy instrumentation and advanced machinery to meet diverse consumer needs

What does the company do to achieve accuracy and precision?
It closely collaborates with microcomputer machinery manufacturers and machinery calibrators.

1. _____

   _____

2. _____

   _____

3. _____

   _____

K Based on the three situations in this unit, write three questions beginning with *How*, and answer them. The questions do not need to come directly from these situations.

Examples

How long has cancer been the leading cause of death in Taiwan?

Since 1982

How long has the Cancer Center of National Taiwan University Hospital been operating?

Since 1999

1. _____

   _____

2. _____

   _____

3. _____

   _____

L Based on the three situations in this unit, write three questions beginning with **Why**, and answer them. The questions do not need to come directly from these situations.

Examples

Why has the Radiology Technology Department at Cheng Hsin Medical Center gained much experience?

It has trained many outstanding radiologists who are now spread throughout Taiwan

Why does the radiology unit have state-of-the-art facilities and highly skilled personnel?

To provide prompt and efficient care

1. _____

   _____

2. _____

   _____

3. _____

   _____

M Write questions that match the answers provided.

_____

_____

in 1967

_____

_____

Into radiology, nuclear medicine and radiotherapy sections, specializing in x-ray, CT and MRI examinations, respectively

_____

_____

Through its state-of-the-art facilities and highly skilled personnel

N Listening Comprehension II

Situation 4

1. How do group leaders coordinate the efforts of employees in their groups?

   A. by adopting the latest technological practices and stressing workplace safety

   B. by university graduates with backgrounds in quality control and related fields

   C. by encouraging strong cooperation through constant training to drill members on independent problem solving

2. How does the Administrative Department emphasize product quality?

   A. by adopting the latest technological practices and stressing workplace safety

   B. by emphasizing customer needs to ensure consumer expectations are satisfied

   C. by adopting advanced technologies to gain consumer confidence

3. What requires a thorough understanding of design quality and quality control to maintain sound manufacturing practices?

   A. extensive experience in developing high-accuracy instrumentation and advanced machinery

   B. a global perspective towards manufacturing quality products

   C. compliance with ISO 14000 standards

4. How many employees comprise the administrative staff?

   A. 25

   B. 30

   C. 40

5. Why does the Administrative Department attempt to rigorously control manufactured product quality?

   A. to ensure consumer expectations are satisfied

   B. to gain consumer confidence

   C. to satisfy consumer demand

Situation 5

1. What has the enhanced form of brachytherapy offered at the Cancer Center achieved?

   A. 5-year survival in 95% of patients with superficial tumors

   B. 5-year survival in 95% of patients with malignant tumors

   C. 5-year survival in 95% of patients with advanced pharyngeal tumors

2. When was the Cancer Center of National Taiwan University Hospital established?

   A. in 1982

   B. in 1991

   C. in 1999

3. In what aspects of cancer research does the Cancer Center nurture professional talent?

   A. brachytherapy for cervical, ovarian and bladder cancers

   B. surgical oncology, radiation oncology and chemotherapy

   C. therapeutic, physiological and psychological

4. How long has cancer been the leading cause of death in Taiwan?

   A. since 1999

   B. since 1982

   C. since 1985

5. What does the Cancer Center's customized treatment strategy incorporate?

   A. the latest treatments in surgical oncology, radiation oncology and chemotherapy

   B. innovative research to prevent, detect and treat various cancers in their early stages

   C. the enhanced form of brachytherapy

Situation 6

1. How many decades of services has the Radiology Technology Department at Cheng Hsin Medical Center provided?

   A. more than five decades

   B. nearly four decades

   C. nearly five decades

2. Why does the radiology unit have state-of-the-art facilities and highly skilled personnel?

   A. to operate a residential training program under the auspices of the Department of Health

   B. to provide prompt and efficient care

   C. to provide imaging and functional evaluations for various body organs

3. How is the Medical Imagery Department divided?

   A. into radioimmunoassay analysis, nuclear cardiology and neuropsychiatry

   B. into radiology, nuclear medicine and radiotherapy sections

   C. into high-tech magnetic resonance imaging system, rapid computerized tomography, a dual energy x-ray bone densitometer, mammography, high-resolution ultrasonography and digital subtraction angiography

4. Why does the radiology team continuously pursue advanced technology applications?

   A. to offer continually improving therapeutic services

   B. to accurately detect tumor size and location

   C. to receive accreditation from the Chinese Society of Radiology

5. Why is a research project currently underway on administering radioactive iodine 13 1?

   A. digitalizing all of the hospital's medical images

   B. to determine accurate dosages for therapeutic purposes

   C. to treat patients with thyroid disorders as a part of long-term follow-up therapy

211

O Reading Comprehension II
Pick the work or expression whose meaning is closest to the meaning of the underlined word or expression in the following passages.

Situation 4

1. Established in 1988 as a producer of quality motors, Yuan Electrical Machinery Company has <u>accumulated</u> extensive experience in developing high-accuracy instrumentation and advanced machinery to meet diverse consumer needs.

   A. dissipate

   B. amass

   C. disperse

2. The company closely collaborates with microcomputer machinery manufacturers and machinery calibrators to achieve accuracy and <u>precision</u>.

   A. ambiguity

   B. approximation

   C. exactitude

3. Additionally, the company adopts a closely supervised quality control inspection system in manufacturing to <u>ensure</u> that its motors have high torsion, low noise emissions and long life.

   A. guarantee

   B. nullify

   C. traverse

4. Machinery is <u>exported</u> mainly to Germany, Hong Kong, Japan, the United States and other southeast Asian countries.

   A. brought into the country

   B. sent from abroad

C. shipped overseas

5. Led by a chief administrator in charge of quality control, an administrative staff of 30 employees focuses on five to six quality control-<u>related</u> items.

    A. pertinent

    B. irrelevant

    C. moot

6. Group leaders <u>coordinate</u> the efforts of employees in their groups by encouraging strong cooperation through constant training to drill members on independent problem solving skills.

    A. muddle

    B. confound

    C. collocate

7. The Administrative Department is mainly staffed by university graduates with backgrounds in quality control and related fields, with several also having masters and doctoral degrees in fields related to <u>enhancing</u> product quality.

    A. depraving

    B. debasing

    C. elevating

8. The Department <u>recruits</u> primarily from nearby universities and technological institutes, offering recent university graduates numerous management opportunities.

    A. incenses

    B. conscripts

    C. displeases

9. While adopting a global perspective towards manufacturing quality products, the department focuses on the following directions: <u>rigorously</u> controlling manufactured product quality to satisfy consumer demand, adopting advanced technologies to gain consumer confidence, emphasizing customer needs to ensure consumer expectations are satisfied and approaching management with a sense of community service.

   A. flexibly

   B. leniently

   C. austerely

10. Particularly, the Department emphasizes product quality by adopting the latest technological practices and stressing workplace <u>safety</u>.

    A. invulnerability

    B. perils

    C. predicaments

11. Demonstrating its commitment to excellence, the department contributed <u>crucially</u> to the company receiving ISO 9000 and 9002 product certification.

    A. unequivocally

    B. inconsequentially

    C. evanescently

12. Furthermore, the Department is striving to achieve <u>compliance with</u> ISO 14000 standards, something which requires a thorough understanding of design quality and quality control to maintain sound manufacturing practices.

    A. antipathy towards

    B. objection to

    C. adherence to

Situation 5

1. Committed to providing advanced research <u>capabilities</u> and expertise in cancer therapy, the Cancer Center of National Taiwan University Hospital was established in 1999, and comprises a chemotherapy department, radiation therapy department, nuclear medicine laboratory, clinical experiment laboratory, radiology biological laboratory, biological statistical laboratory, outpatient services for tumor victims, a chemotherapy treatment room and a ward for tumor patients.

   A. inaptitude

   B. forte

   C. inefficacy

2. The Radiation Therapy Department is the <u>heart</u> of the Cancer Center and focuses on four treatment areas: a state-of-the-art IMRT-linear accelerator, Co-60 therapeutic machinery, a computer tomography simulator and a block-making room.

   A. fringe

   B. periphery

   C. center

3. While providing a diverse range of services for cancer patients, the Cancer Center provides a <u>customized</u> treatment strategy that incorporates the latest treatments in surgical oncology, radiation oncology and chemotherapy.

   A. uniform

   B. conventional

   C. tailored

4. The center has a highly <u>talented</u> staff, including radiation oncologists, surgical oncologists, medical oncologists, diagnostic radiologists, that operates advanced instrumentation such as a linear accelerator, CO-60 and CT, as well as medical physicists who plan therapy.

A. adroit

B. vacillating

C. capricious

5. These talented staff help the Cancer Center to achieve the following objectives: providing patient-focused healthcare to enhance patient quality of life, satisfying individual patient requirements and those of their relatives in a nurturing environment, fostering innovative research to prevent, <u>detect</u> and treat various cancers in their early stages and training future cancer treatment experts and researchers.

A. neglect

B. overlook

C. pinpoint

6. Additionally, while conducting <u>preliminary</u> and clinical research, the Cancer Center nurtures professional talent in therapeutic, physiological and psychological aspects of cancer research.

A. terminating

B. inaugural

C. inevitable

7. Moreover, nurses receive specialized training to care for radiation therapy patients, including help such patients cope with the discomfort caused by <u>adverse</u> skin effects and gland deficiencies.

A. afflictive

B. benign

C. auspicious

8. The Cancer Center <u>focuses</u> particularly on breast, nasal and pharyngeal, and lung cancers.

   A. diverts

   B. sidetracks

   C. converges

9. The center also <u>provides</u> brachytherapy for cervical, ovarian and bladder cancers.

   A. confines

   B. restrains

   C. furnishes

10. Notably, the enhanced form of brachytherapy offered at the center achieves 5-year survival in 95% of patients with <u>superficial</u> tumors.

    A. cavernous

    B. shallow

    C. extensive

11. Besides providing state-of-the-art medical services for patients, the Center also offers support for patients and relatives <u>coping with</u> cancer therapy.

    A. contravening

    B. grappling with

    C. disavowing

Situation 6

1. Established in 1967, the Radiology Technology Department at Cheng Hsin Medical Center has provided nearly five decades of service, and has trained many outstanding radiologists who are now <u>spread</u> throughout Taiwan.

   A. authorized

   B. obliged

C. disseminated

2. The Medical Imagery Department is <u>divided</u> into radiology, nuclear medicine and radiotherapy sections, specializing in x-ray, CT and MRI examinations, respectively.

A. synthesized

B. consolidated

C. segmented

3. The radiology unit has state-of-the-art facilities and highly skilled personnel to provide <u>prompt</u> and efficient care.

A. expeditious

B. lethargic

C. dilatory

4. The radiology team continuously <u>pursues</u> advanced technology applications to offer continually improving therapeutic services.

A. flies the coop

B. absconds

C. perseveres in

5. With its excellent specialists and facilities, the radiology unit has not only received accreditation from the Chinese Society of Radiology, but also operates a residential training program under the <u>auspices</u> of the Department of Health.

A. admonition

B. sanction

C. remonstration

6. The specialized equipment of the unit <u>includes</u> a high-tech magnetic resonance imaging system, rapid computerized tomography, a dual energy x-ray bone densitometer, mammography, high-resolution ultrasonography and digital subtraction angiography.

  A. repudiates

  B. encompasses

  C. ostracize

7. Nuclear medicine provides imaging and functional <u>evaluations</u> for various body organs, as well as radioimmunoassay analysis, including nuclear cardiology and neuropsychiatry.

  A. reprimands

  B. devaluations

  C. appraisals

8. A research project is currently underway on administering radioactive iodine 13 1 to <u>treat</u> patients with thyroid disorders as a part of long-term follow-up therapy.

  A. cure

  B. inflict

  C. wreak

9. The radiotherapy unit is fitted with a double density linear accelerator for treating patients suffering from cancer, supported by a high precision radiographic simulator, for <u>accurately</u> detecting tumor size and location.

  A. precisely

  B. haphazardly

  C. erroneously

10. <u>Driven</u> by a high-powered computer, the therapeutic planning system can determine accurate dosages for therapeutic purposes.

    A. detracted

    B. impelled

    C. deflected

11. This service prioritizes <u>quality</u> control in therapeutic planning.

    A. void

    B. attribute

    C. chasm

12. As for future trends, the hospital is <u>digitalizing</u> all of its medical images, and the PACS system will provide the entire hospital with an image transmissions system in June of this year.

    A. manually recording

    B. computerizing

    C. drafting

Unit Six

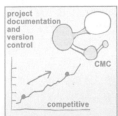 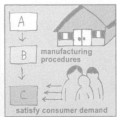 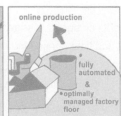

 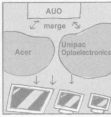 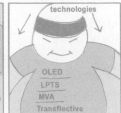

Introducing a Technology

科技介紹

1. Briefly explain the factors (internal and external) factors that influence development of this technology in Taiwan.
   影響台灣科技發展的因素

2. Point out the unique characteristics of this technology development in Taiwan.
   科技發展的特色

3. List the objectives of how to further develop this technology.
   發展科技的目標

4. Define the role of this technology in relation to environmental, manufacturing or technology problems.
   界定科技的地位：扼要解釋在環境、製造或科技方面的問題

5. List applications of this technology made so far, highlighting any particular characteristics or features that are unique to Taiwan's circumstances.
   目前從事的應用科技：(1)應用科技的特色或特徵，(2)特殊的個案；科技市場的契機；繼續從事科技應用的未來挑戰

Vocabulary and related expressions

enhancing its competitive edge	強化其競爭優勢
inefficient and time consuming	無效率以及花時間的
production bottlenecks	生產瓶頸
making on-line queries	提出線上質問
global competitiveness	全球性的競爭
elevated living standards	提升生活水準
strong consumer demand	強烈的消費者需求
prohibitive costs	過高的價格
phenomenal growth	驚人的成長
sustained growth	持續的成長
greater likelihood of	有更大的……可能性
most extensively adopted	最被廣泛地採用
complying with legal standards	遵從合法標準
technological constraints	科技的約束
overcoming barriers	克服障礙
pervades daily life	遍及於每日生活
to remain abreast of	保持與……並列
diversifying	增加產品種類以擴大（業務）
professional integrity	專業的正直
environmental consciousness	環境意識
cultivating social consciousness	培育社會意識
increased range and firepower	增加範圍幅度及火力

Situation 1

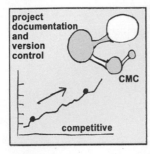

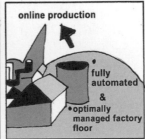

Situation 2

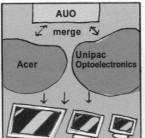
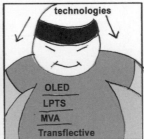

Situation 3

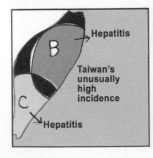
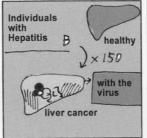
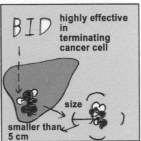

A Write down the key points of the situations on the preceding page, while the instructor reads aloud the script from the Answer Key. Alternatively, students can listen to the script online at www.chineseowl.idv. tw

Situation 1

_____

_____

Situation 2

_____

_____

_____

Situation 3

_____

_____

_____

B Oral practice I

Based on the three situations in this unit, write three questions beginning with *How*, and answer them. The questions do not need to come directly from these situations.

Examples

How did China Motors Corporation (CMC) increase its productivity and enhance its competitive edge in the domestic automotive sector?

By constructing a project documentation and version control (PDVC) system

How was production of new automobiles an inefficient and time consuming process.

They were spray painted on the factory floor.

1. _____

_____

2. _____

_____

3. _____

_____

C Based on the three situations in this unit, write three questions beginning with *What*, and answer them. The questions do not need to come directly from these situations.

Examples

What has created added value and increased efficiency?

Product innovation

What could not simultaneously achieve a competitive price, large size and high quality until the 1990s?

Liquid crystal displays

1. _____

   _____

2. _____

   _____

3. _____

   _____

D Based on the three situations in this unit, write three questions beginning with **Why**, and answer them. The questions do not need to come directly from these situations.

Examples

Why are there high levels of hepatitis associated liver cancer among Taiwanese?

Because individuals with Hepatitis B have a 150 times greater likelihood of contracting liver cancer than those without the virus

Why has liver cancer ranked as a leading cause of mortality in Taiwan during the past decade?

Because of Taiwan's unusually high incidence of Hepatitis B and Hepatitis C

1. _____

_____

2. _____

_____

3. _____

_____

E Write questions that match the answers provided.

1. _____

   _____

   Liver cancer

2. _____

   _____

   Over 10,000

3. _____

   _____

   Five

F Listening Comprehension I

Situation 1

1. How do global automakers generally produce cars?

    A. in large quantities

    B. in smaller quantities

    C. with more diversification

2. How does the advanced PDVC system improve automation control technologies?

    A. by focusing mainly on flexibility in production scheme, online production and further integration of information technologies

    B. by producing a smaller quantity of more diverse products

    C. by foreseeing potential production bottlenecks

3. How much has the PDVC system saved China Motors Corporation?

    A. NT$ 2,000,000 over the past six months

    B. NT$ 2,000,000 over the past year

    C. NT$ 2,000,000 over two years

4. How does the PDVC system help to simultaneously optimize factory floor use and satisfy consumer demand?

    A. by arranging the simultaneous production of various models, striving to fill factory orders, and making on-line queries regarding how to increase upper and lower supply chain efficiency

    B. by eliminating an inefficient and time consuming process

    C. by producing a smaller quantity of more diverse products than global automakers do

5. How has China Motors Corporation (CMC) increased its productivity and enhanced its competitive edge in the domestic automotive sector?

    A. by arranging the simultaneous production of various models, striving to fill factory orders, and making on-line queries regarding how to increase upper and lower supply chain efficiency

B. by recently constructing a project documentation and version control (PDVC) system

C. by simultaneously optimizing factory floor use and satisfying consumer demand

Situation 2

1. Why has AU Optronics Corporation not yet developed the equivalent technology for extremely large sized displays?

   A. the inability to provide both large and small-medium applications

   B. the inability to simultaneously achieve a competitive price, large size and high quality

   C. primarily owing to the prohibitive costs

2. What has the ability of AU Optronics Corporation to integrate life sciences and medical technology resulted in?

   A. the effective integration of multidisciplinary fields

   B. state-of-the-art photoelectronic laser technology applications

   C. technological advances in manufacturing liquid crystal displays

3. How many employees did AU Optronics Corporation have by the end of 2003?

   A. over 20,000

   B. nearly 20,000

   C. 20,000

4. How many local and international patents does AU Optronics Corporation own?

   A. 656

   B. 676

   C. 556

5. What will enable AU Optronics to continue to succeed?

   A. the ability to produce large-sized panels for use in desktop monitors, notebook

PCs, LCD TVs and other consumer appliances

B. the most advanced TFT-LCD technologies

C. the effective integration of multidisciplinary fields such as optoelectronics, mechanical engineering, electrical engineering and materials science

Situation 3

1. What is the most extensively adopted of the following therapeutic methods?

    A. microwave therapy

    B. high temperature therapy

    C. radiation therapy

2. Why has liver cancer ranked as a leading cause of mortality in Taiwan during the past decade?

    A. because of Taiwan's unusually high incidence of Hepatitis B and Hepatitis

    B. owing to the inability to reduce damage to normal cells

    C. owing to the inability to treat various cancers using radiation therapy

3. What explains the high levels of hepatitis associated liver cancer among Taiwanese?

    A. Taiwan has over 10,000 fatalities annually from chronic liver hepatitis, cirrhosis or liver cancer.

    B. 80 to 90% of those affected carry the Hepatitis B or Hepatitis C viruses.

    C. Individuals with Hepatitis B have a 150 times greater likelihood of contracting liver cancer than those without the virus.

4. How has radiation therapy advanced rapidly during the past decade?

    A. from microwave therapy to high temperature therapy

    B. from CO-60 to 3D conformal therapy

    C. from radiation therapy to high temperature therapy

5. Why do patients treated with BID receive irradiation treatment in the morning?

    A. to integrate BID with other therapeutic methods

    B. to damage the DNA of cancer cells

    C. to reduce damage to normal cells

G Reading Comprehension I
Pick the work or expression whose meaning is closest to the meaning of the underlined word or expression in the following passages.

Situation 1

1. China Motors Corporation (CMC) recently constructed a project documentation and version control (PDVC) system, thus increasing its <u>productivity</u> and enhancing its competitive edge in the domestic automotive sector.

   A. atrophy

   B. ingenuity

   C. retrogression

2. Previously, new automobiles were spray painted on the factory floor, an <u>inefficient</u> and time consuming process.

   A. potent

   B. inept

   C. puissant

3. The advanced PDVC system improves automation control technologies by <u>foreseeing</u> potential production bottlenecks.

   A. recalling

   B. reminiscing

   C. prognosticate

4. Whereas global automakers generally produce cars in large quantities, CMC produces a smaller quantity of more <u>diverse</u> products.

   A. heterogeneous

   B. uniform

   C. homogeneous

5. Demonstrating its success, the PDVC system has reduced personnel numbers on the factory floor from fifteen to six, while reducing <u>wasted</u> storage capacity from 68,071 to 55,031 units, saving CMC NT$ 2,000,000 over two years.

   A. conserved

   B. sustained

   C. expended

6. <u>Catering</u> to individualized consumer tastes requires the automobile industry to effectively manage manufacturing procedures to simultaneously optimize factory floor use and satisfy consumer demand.

   A. repulsing

   B. coddling

   C. impeding

7. Catering to individualized consumer tastes requires the automobile industry to effectively manage manufacturing procedures to simultaneously <u>optimize</u> factory floor use and satisfy consumer demand.

   A. maximize

   B. attenuate

   C. curtail

8. The PCVC system helps to achieve this by <u>arranging</u> the simultaneous production of various models, striving to fill factory orders, and making on-line queries regarding how to increase upper and lower supply chain efficiency.

   A. facilitating

   B. hindering

   C. inhibiting

9. Such technological innovations will <u>ultimately</u> improve the global competitiveness of the Taiwanese automotive sector.

   A. originally

B. initially

C. eventually

10. Technological developments in this sector focus mainly on <u>flexibility</u> in production scheme, online production and further integration of information technologies and, ultimately, result in a fully automated and optimally managed factory floor.

A. rigidity

B. pliability

C. austerity

Situation 2

1. Elevated living standards and strong consumer demand in Taiwan have <u>fueled</u> technological advances in manufacturing liquid crystal displays.

A. suppressed

B. quelled

C. ignited

2. Despite having successfully developed technology for producing medium and full sized liquid crystal color displays, AU Optronics Corporation has yet to develop the <u>equivalent</u> technology for extremely large sized displays, primarily owing to the prohibitive costs.

A. commensurate

B. asymmetrical

C. lopsided

3. Until the 1990s, liquid crystal displays could not <u>simultaneously</u> achieve a competitive price, large size and high quality.

A. exclusively

B. concomitantly

C. individually

4. The main applications of liquid crystal displays are in notebooks and cellular phones. AU Optronics Corporation mainly produces large-sized panels for use in desktop monitors, notebook PCs, LCD TVs and other consumer <u>appliances</u>.

A. software

B. furnishings

C. online security devices

5. The company's retail sales of such panels reached NT$14.596 million for March 2005, with shipments reaching 2.33 million <u>pieces</u> during the same month, increased 29.1% from the previous month.

A. measurements

B. dollars

C. units

6. Further demonstrating the <u>phenomenal</u> growth of AU Optronics recently, shipments of small- and medium-sized panels increased by 23.4% to reach 3.39 million units during March 2005.

A. prodigious

B. inferior

C. substandard

7. With the <u>merger</u> of Acer Display Technology Corporation (a subsidiary of Acer) and Unipac Optoelectronics Corporation (a subsidiary of United Microelectronics Corporation) in 2001, AU Optronics obtained the most advanced TFT-LCD technologies and became known as a one-stop-shop capable of providing both large and small-medium applications.

A. insulation

B. segregation

C. conglomeration

8. By the end of 2003, the operations of AU were spread across China, Europe, Japan, South Korea, Taiwan and the United States, the company had over 20,000 employees, and annual revenues had <u>exceeded</u> $US 3 billion.

A. rescinded

B. surpassed

C. abrogated

9. Notably, the ability of the corporation to <u>integrate</u> life sciences and medical technology has resulted in state-of-the-art photoelectronic laser technology applications.

A. homogenize

B. differentiate

C. distinguish

10. Given its technological capabilities, AU Optronics is <u>forging</u> ahead into OLED, LPTS, MVA and transflective technologies.

A. vacating

B. recoiling

C. plodding

11. Product innovation has created added value and increased efficiency, leading to <u>sustained</u> growth in corporate profits and enhancing TFT-LCD performance, as demonstrated by the 656 local and international patents the company owns.

A. faltering

B. wavering

C. persistent

12. The effective integration of multidisciplinary fields such as optoelectronics, mechanical engineering, electrical engineering and materials science will enable AU Optronics to continue to <u>succeed</u>.

A. flourish

B. flounder

C. blunder

Situation 3

1. Liver cancer has ranked as a leading cause of mortality in Taiwan during the past decade because of Taiwan's unusually high <u>incidence</u> of Hepatitis B and Hepatitis C.

A. infrequency

B. prevalence

C. rarity

2. Taiwan has over 10,000 fatalities annually from <u>chronic</u> liver hepatitis, cirrhosis or liver cancer, and 80 to 90% of those affected carry the Hepatitis B or Hepatitis C viruses.

A. transient

B. deep-rooted

C. fugacious

3. Individuals with Hepatitis B have a 150 times greater <u>likelihood</u> of contracting liver cancer than those without the virus, explaining the high levels of hepatitis associated liver cancer among Taiwanese.

A. entanglement

B. quandary

C. verisimilitude

4. Five major <u>therapeutic</u> methods are available for treating liver cancer: surgery, including either therapeutic surgery or organ transplantation; blockage of blood flow, such as embolism of the liver artery; chemotherapy, including treatment for the entire body and interarterial injection; local injection, including injection of alcohol and glacial acetic acid; and temperature therapy, including microwave

therapy, high temperature therapy and radiation therapy.

A. recuperative

B. debilitative

C. enervative

5. Radiation therapy is the most <u>extensively</u> adopted of these therapeutic methods.

A. commodiously

B. scantily

C. meagerly

6. <u>Advances in</u> digital technology have led to 3D computer-animated control stations being widely adopted to treat various cancers using radiation therapy.

A. relapses

B. ascension

C. recidivism

7. Radiation therapy has advanced rapidly during the past <u>decade</u>: from CO-60 to 3D conformal therapy.

A. biennially

B. annually

C. ten years

8. Previously, surgery was the only therapeutic method of <u>treating</u> liver cancer.

A. exacerbating

B. coping with

C. aggravating

9. While the one-year <u>survival</u> rate following surgery can reach 80%, the five year survival rate is only 50%.

A. sustenance

B. mortality

C. fatality

10. Ultimately, survival rates depend on <u>factors</u> such as tumor position, size, degree of cirrhosis and meta situation.

    A. hazards

    B. perils

    C. considerations

11. However, only 30% of liver cancer patients are suitable surgery candidates because hepatitis B and hepatitis C worsen the <u>prognosis for</u> cancer meta. Still, surgery is desirable for alleviating the adverse effects of irradiation and so enabling the radiation dose to be increased to 6600rad, thus achieving survival rates equaling those of conventional surgical therapy.

    A. result of

    B. consequence of

    C. likelihood of

12. Regular treatment is an <u>essential</u> part of liver cancer therapy, and patients must be treated twice daily.

    A. intrinsic

    B. immaterial

    C. trivial

13. Such regular treatment can completely <u>cure</u> over 60% of sufferers.

    A. defile

    B. restore

    C. mar

14. A novel therapeutic method, BID, attacks liver cancer cells with nearly twice the <u>intensity</u> of conventional approaches.

    A. attenuation

    B. magnitude

    C. delicacy

15. Patients treated with BID receive irradiation treatment in the morning to damage the DNA of cancer cells; a <u>subsequent</u> treatment six to eight hours later further damages the DNA.

A. ensuing

B. precursory

C. anterior

16. BID is highly effective in terminating cancer cells but has certain <u>limitations</u>: the liver cancer must be smaller than 5 cm; cancer cells will still be present after surgery; and liver cancer tumors larger than 5-8 cm can not be treated.

A. advantages

B. merits

C. impediments

17. BID photon knife therapy has high <u>efficacy</u> and reduces damage to normal cells.

A. ineptitude

B. fragility

C. potency

18. Integrating BID with other therapeutic methods, including embolism, alcohol <u>injection</u>, surgery or chemotherapy, will ultimately reduce the incidence of liver cancer in Taiwan.

A. retraction

B. inoculation

C. rescission

H Common elements in introducing a technology科技介紹 include the following elements:

1. Briefly explain the factors(internal and external) factors that influence development of this technology in Taiwan.
影響台灣科技發展的因素

· The increasing market value of biotechnology products in Taiwan cannot be underestimated, with its development underway since the 1980s. However, despite the considerable amount of resources and efforts committed to its development island wide, few noteworthy achievements have been made. Given its high creativity value and attractive marketing features, biotechnology will generate large revenues once products are commercialized.

· 3D automation technology is the driving force behind Taiwan's online gaming sector, which is forecasted to have a market value of $NT 533,000,000. Therefore, the domestic on-line gaming sector must continuously acquire new technological capabilities that are in line with the global market demand to continually refine the quality of on-line games.

· Hospitals devote a significant amount of their annual operating budgets to upgrading their information network capacity through the purchase of advanced hardware and software. Adopting the latest information technologies can greatly enhance a hospital's administrative and managerial capabilities.

2. Point out the unique characteristics of this technology development in Taiwan.
科技發展的特色

· Whereas biotechnology was seldom researched in Taiwan just a few years ago, this multidisciplinary field is now actively researched island wide, as evidenced by its use in plants and animals for agricultural production.

· Through the encouragement of Taiwan's National Health Insurance Bureau, many hospitals are incorporating a medical information system (MIS) in daily operations to increase the efficiency of physicians in examining patients, diagnosing their illnesses and prescribing therapeutic treatment.

· Since the Taiwanese government prioritized biotechnology as a major area of national development in 1982, many industrial sectors have actively engaged in related projects synergistically.

3. List the objectives of how to further develop this technology.
發展科技的目標

· State-of-the-art design technology in Elsa's line of products leaves customers with the impression that each mattress is suited to a particular user's tastes and comfort needs.

· By adopting basic life sciences as a tool to understand many mysteries of life, biotechnology integrates the efforts of seemingly polar disciplines such as medicine, cosmetics, agriculture, animal husbandry, spinning and weaving, oceanic studies and national defense.

· Despite the pervasiveness of biosensors in various fields, their role in health care monitoring can not be underestimated, especially given the need for continuous monitoring of analytes such as glucose and urea. As invasive biosensing methods in self-testing kits frequently cause patient discomfort, non-invasive testing using sweat, saliva or skin samples has become increasingly popular.

4. Define the role of this technology in relation to environmental, manufacturing or technology problems.
界定科技的地位：扼要解釋在環境、製造或科技方面的問題

· Eliminating lead found in medical waste using advanced treatment technologies is of priority concern given the storage of waste in airtight metal container, in which melting occurs during incineration between 1500°C- 3000°C.

· According to statistics from Taiwan's Environmental Protection Administration, 60 million CD ROMs are discarded annually Additionally, incinerating discarded CD-ROM produces dioxin, posing a potential environmental threat. As CD ROMs store historical, library, governmental , commercial and personal datatheir convenience and benefits are undisputed. However, given their adverse environmental impact, the Taiwanese government should encourage their reuse to avoid pollution by establishing retrieval centers to conserve resources through recovery.

· With national development strategies heavily focused on the output value on the island's medical sector, the Taiwanese government is in the process of developing medical health care scheme that, in addition to closely inspecting the health services of domestic medical institutions, aims to devise and implement a nationwide health information system by 2003.

5. List applications of this technology made so far, highlighting any particular characteristics or features that are unique to Taiwan's circumstances.

目前從事的應用科技：(1) 應用科技的特色或特徵，(2) 特殊的個案；科技市場的契機；繼續從事科技應用的未來挑戰

· Besides enhancing the frame quality with advanced animation capabilities, 3D automation technology also reduces overhead costs in game development.

· Advanced treatment technologies such as the MESSCUD system ensure that the above problem is resolved owing to the system's following features: relative ease in waste management without the requirement of classifying waste, use of a unique storage container, ability to effectively handle potential lead problems caused by high-temperature melting, recycling after incineration and no risk of infection following treatment.

· Market investment strategies now heavily emphasize the ability to concentrate governmental resources in the hi-tech sector in order to encourage product innovativeness.

In the space below, introduce a technology.

_____

_____

_____

_____

_____

_____

_____

_____

_____

Look at the following examples of introducing a technology.

Technological advances in manufacturing bed mattresses are largely guided by elevated living standards in Taiwan and a strong consumer demand for a restful night's sleep. Already a patented technology, the Elsa mattress brand offers not only full body support through seven regions, i.e., the head, neck, shoulders, back, waist, buttocks and legs, but also two softer comfort zones and one coil spring zone, allowing consumers to sleep more peacefully at night. The softer textured material enables individuals to rest in a more natural posture, as supported by coil springs that match the body's contour shape. Additionally, Elsa's commitment to mattress durability is reflected in the product's ten year guarantee. State-of-the-art design technology in Elsa's line of products leaves customers with the impression that each mattress is suited to a particular user's tastes and comfort needs. For instance, measurement machinery imported from Germany ensures that the mattress corresponds to the user's physical dimensions so that maximum comfort is ensured. One concern, however, is that customers accustomed to sleeping on a pullout couch bed may find an Elsa mattress too soft. Elsa is currently devising an approach to effectively address this issue. Given its above success in providing customer satisfaction, Elsa continues to remain abreast of the latest technological trends in sleeping comfort in order to maintain a competitive edge in this growth sector in Taiwan.

The increasing market value of biotechnology products in Taiwan cannot be underestimated, with its development underway since the 1980s. However, despite the considerable amount of resources and efforts committed to its development island wide, few noteworthy achievements have been made. Given its high creativity value and attractive marketing features, biotechnology will generate large revenues once products are commercialized. Since the Taiwanese government prioritized biotechnology as a major area of national development in 1982, many industrial sectors have actively engaged in related projects synergistically. Despite the infancy of this industry domestically, biotechnology integrates a diverse array of multidisciplinary fields. Following two decades of development globally, biotechnology continues to advance at an accelerated pace. By adopting basic life sciences as an effective means of understanding many mysteries of life, biotechnology integrates the efforts of seemingly polar disciplines such as medicine, cosmetics, agriculture, animal husbandry, spinning and weaving, oceanic studies and national defense. Given growing public concern over individual appearances, many cosmetic products have been developed locally or imported from abroad. Bio-cosmetics is thus a highly profitable area that biotechnology has recently focused on through the use of natural plant extracts. To become competitive in this intensely fierce global arena, local manufacturers must implement appropriate management strategies in order to gain a larger share of the domestic market.

3D automation technology is the driving force behind Taiwan's online gaming sector, which is forecasted to reach a market value of $NT 533,000,000. Therefore, the domestic on-line gaming sector must continuously acquire new technological capabilities that are in line with the global market demand. The demand for 3D automation technology arises from the need to continually refine the quality of on-line games. Besides enhancing the frame quality with advanced animation capabilities, 3D automation technology also reduces overhead costs in game development. More than for use in the online gaming sector, 3D automation technology is also adopted in console games, personal computer games and arcade games.

Situation 4

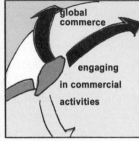

Situation 5

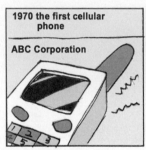
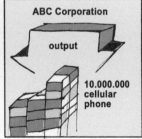

Situation 6

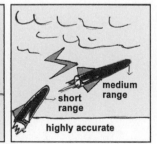

I Write down the key points of the situations on the preceding page, while the instructor reads aloud the script from the Answer Key. Alternatively, students can listen to the script online at www.chineseowl.idv. tw

Situation 4

_____

_____

Situation 5

_____

_____

_____

Situation 6

_____

_____

_____

J Oral practice II

Based on the three situations in this unit, write three questions beginning with **What**, and answer them. The questions do not need to come directly from these situations.

Examples

What makes it impossible for enterprises to integrate their efforts with other manufacturers when complying with legal standards and attempting to better understand consumer behavior?

Inefficient computerization

What faces challenges in transaction and distribution?

Local e-commerce

1. _____

   _____

2. _____

   _____

3. _____

   _____

K Based on the three situations in this unit, write three questions beginning with *How*, and answer them. The questions do not need to come directly from these situations.

Examples

How has ABC Corporation become the largest global cellular phone manufacturer?

By diversifying from advanced manufacturing technologies into specialized design processes

How did ABC Corporation generate revenues of US$ 1,500,000,000 last year?

With an annual output of 10,000,000 cellular phones

1. _____

   _____

2. _____

   _____

3. _____

   _____

L Based on the three situations in this unit, write three questions beginning with **Why**, and answer them. The questions do not need to come directly from these situations.

Examples

Why does the Ting-Kung 2 missile have increased range and firepower?

By adopting an active radar homing seeker

Why was the Ray-Ting 2000 designed?

To repel an amphibious attack and provide superior firepower to conventional tube artillery

1. _____

_____

2. _____

_____

3. _____

_____

M Write questions that match the answers provided.

_____

_____

_____

Manufacturing ground-to-air guided anti-missile systems

_____

_____

_____

Shiung-Feng 1 and 2

_____

_____

_____

The Ray-Ting 2000

N Listening Comprehension II

Situation 4

1. Why must the workplace and even global commerce, governments and enterprises worldwide invest heavily in electronic commerce?

 A. to increase national GDP and corporate competitiveness

 B. to integrate Internet-based services through technical support, marketing strategies, market positioning and collaboration with partners

 C. to boost global commerce

2. What makes it impossible for enterprises to integrate their efforts with other manufacturers when complying with legal standards and attempting to better understand consumer behavior?

 A. the inability to ensure network security for online purchases

 B. difficulties in transaction and distribution

 C. inefficient computerization adopted by many enterprises

3. Why are customers hesitant to use credit cards online?

 A. difficulties in transaction and distribution

 B. hackers

 C. obstacles to fully developing e-commerce

4. How can enterprises overcome barriers and engage in commercial activities as soon as possible?

 A. by promoting brand image in emerging markets

 B. through accelerated development of e-commerce

 C. by ensuring network security for online purchases

5. Why did the Taiwanese government draft and legislate an electronic commerce policy in 1998?

    A. to reduce barriers to market entry and increase returns on investment

    B. to integrate Internet-based services through technical support, marketing strategies, market positioning and collaboration with partners

    C. to remain abreast of global trends in e-commerce and protect the future standing of Taiwan throughout the Asian-Pacific region

Situation 5

1. What must global enterprises commit themselves to?

    A. integrating electronic equipment, silicon chips, chemical processes, gas water and electricity

    B. professional integrity and environmental consciousness

    C. satisfying consumer demand with quality technological products that have low power consumption and comprehensive technical services

2. How has ABC Corporation tried to be a sustainable and environmentally responsible company?

    A. by cultivating social consciousness

    B. by creating employment opportunities while operating in a sustainable manner

    C. by maintaining environmentally safe manufacturing practices aimed at protecting employee health and security

3. What should environmentally sound manufacturing processes aim to do?

    A. reduce employee error, conserve water and electricity and reduce pollutant emissions

    B. satisfy consumer demand with quality technological products that have low power consumption and comprehensive technical services

    C. diversify from advanced manufacturing technologies into specialized design processes

4. When did ABC Corporation commercialize the first cellular phone?

   A. in 1970

   B. in 1975

   C. in 1980

5. How much did ABC Corporation generate in revenues last year?

   A. US$ 1,000,000,000

   B. US$ 1,250,000,000

   C. US$ 1,500,000,000

Situation 6

1. What has forced Taiwan to develop its own ground-to-air guided anti-missile system?

   A. to create a balance in the arms buildup involving Taiwan and China

   B. Chinese objections over Taiwanese defense purchases

   C. to repel an amphibious attack and provide superior firepower to conventional tube artillery

2. Where did Taiwan import ground-to-air guided anti-missile systems from?

   A. Canada and the United States

   B. China

   C. its diplomatic allies

3. Who has designed and enhanced the Shiung-Feng 1 and 2, and Ting-Kung 1 and 2 missile systems?

   A. The Taiwanese Ministry of National Defense

   B. Chung Shan Institute of Science and Technology

   C. China

4. How has the Ting-Kung 2 missile increased its range and firepower?

A. by adopting an active radar homing seeker

B. by repelling an amphibious attack and providing superior firepower to conventional tube artillery

C. by coming equipped with a fully automatic fire control system, elevation and azimuth driven systems, as well as positional and direction determining systems

5. What is the Ting-Kung 1 missile fitted with?

A. positional and direction determining systems

B. a fully automatic fire control system

C. a semi-active radar homing seeker

O Reading Comprehension II
Pick the work or expression whose meaning is closest to the meaning of the underlined word or expression in the following passages.

Situation 4

1. Many <u>obstacles</u> exist to fully developing e-commerce in Taiwan.

    A. contingencies

    B. avenues

    C. obstructions

2. Inefficient computerization adopted by many enterprises makes it impossible for enterprises to integrate their efforts with other manufacturers when <u>complying</u> with legal standards and attempting to better understand consumer behavior.

    A. resisting

    B. objecting to

    C. adhering to

3. Local e-commerce also faces <u>challenges</u> in transaction and distribution.

    A. dilemmas

    B. ventures

    C. predicaments

4. For instance, ensuring network security for online purchases is a <u>key</u> concern.

    A. subordinate

    B. paramount

    C. ancillary

5. Hackers remain a concern for consumers and enterprises, making customers <u>hesitant</u> to use credit cards online.

    A. dilatory

B. rejuvenated

C. zealous

6. Technological constraints and cultural considerations mean that the e-commerce sector in Taiwan remains <u>immature</u>.

A. pubescent

B. nubile

C. callow

7. <u>Accelerated</u> development of e-commerce depends on enterprises overcoming barriers and engaging in commercial activities as soon as possible, thus boosting global commerce.

A. expedited

B. deescalated

C. halted

8. Although barriers to market <u>entry</u> are relatively low, returns on investment are high, with potential profits being much higher than in traditional commerce.

A. aperture

B. loophole

C. ingress

9. Additionally, e-commerce can promote brand image in <u>emerging</u> markets.

A. emanating

B. abating

C. subsiding

10. Enterprises must learn to integrate Internet-based services through technical support, marketing strategies, market positioning and <u>collaboration</u> with partners.

A. scuffle

B. hostility

C. connivance

11. The <u>rise</u> of the Internet forces enterprises to adopt the above measures to ensure their survival.

A. abatement

B. ascension

C. mitigation

12. As the Internet <u>pervades</u> daily life, the workplace and even global commerce, governments and enterprises worldwide must invest heavily in electronic commerce to increase national GDP and corporate competitiveness.

A. averts

B. eludes

C. permeates

13. To remain abreast of global <u>trends</u> in e-commerce and protect the future standing of Taiwan throughout the Asian-Pacific region, the Taiwanese government drafted and legislated an electronic commerce policy in 1998.

A. transgression

B. echelon

C. infringement

Situation 5

1. After commercializing the first cellular phone in 1970, ABC Corporation has become the largest global cellular phone manufacturer, <u>diversifying</u> from advanced manufacturing technologies into specialized design processes.

A. varying

B. confining

C. trammeling

2. With an annual output of 10,000,000 cellular phones, ABC Corporation <u>generated</u> revenues of US$ 1,500,000,000 last year.

A. expended

B. squandered

C. spawned

3. Manufacturing cellular phones <u>involves</u> integrating electronic equipment, silicon chips, chemical processes, gas water and electricity.

A. quarantines

B. encompasses

C. precludes

4. The <u>complexity</u> of cellular phones raises environmental protection, security and hygiene issues in manufacturing.

A. simplicity

B. lucidity

C. intricacy

5. To be a sustainable and environmentally responsible company, ABC Corporation has implemented the following measures: maintaining environmentally safe manufacturing practices aimed at protecting employee health and security; <u>strictly</u> adhering to global safety and hygiene practices; educating employees on the importance of environmental protection, security and hygiene during manufacturing; recommending the adoption of novel environmentally friendly and efficient technologies; and extensively testing new materials to reduce risks.

A. compliantly

B. elastically

C. adamantly

6. To be a sustainable and environmentally responsible company, ABC Corporation has implemented the following measures: maintaining environmentally safe

manufacturing practices aimed at protecting employee health and security; strictly adhering to global safety and <u>hygiene</u> practices; educating employees on the importance of environmental protection, security and hygiene during manufacturing; recommending the adoption of novel environmentally friendly and efficient technologies; and extensively testing new materials to reduce risks.

A. putrid

B. sanitary

C. unkempt

7. Global enterprises must commit themselves to professional <u>integrity</u> and environmental consciousness.

A. veracity

B. perverse

C. amoral

8. Environmentally sound manufacturing processes should aim to reduce employee error, conserve water and electricity and reduce pollutant <u>emissions</u>.

A. discharges

B. absorption

C. osmosis

9. Besides simply generating revenues, ABC Corporation must be a responsible community member by <u>cultivating</u> social consciousness, creating employment opportunities while operating in a sustainable manner, seeking to satisfy consumer demand with quality technological products that have low power consumption and comprehensive technical services.

A. eschewing

B. enlightening

C. steering clear of

10. Besides simply generating revenues, ABC Corporation must be a responsible community member by cultivating social consciousness, creating employment opportunities while operating in a sustainable manner and attempting to satisfy consumer demand with quality technological products that have low power consumption and <u>comprehensive</u> technical services.

   A. sweeping

   B. spasmodic

   C. intermittent

Situation 6

1. The Taiwanese Ministry of National Defense has made significant technological <u>advances</u> in manufacturing ground-to-air guided anti-missile systems.

   A. immaterial

   B. nugatory

   C. landmark

2. For instance, besides developing the Ray-Ting 2000 artillery multiple <u>launch</u> rocket system, the Chung Shan Institute of Science and Technology has designed and enhanced the following missile systems: Shiung-Feng 1 and 2, and Ting-Kung 1 and 2.

   A. propel

   B. splash down

   C. descend

3. Previously, Taiwan imported ground-to-air guided anti-missile systems from its diplomatic <u>allies</u>.

   A. assailants

   B. adversaries

   C. confederates

263

4. However, Chinese <u>objections</u> over Taiwanese defense purchases have forced Taiwan to develop its own ground-to-air guided anti-missile system.

A. avows

B. averments

C. demurs

5. Shiung-Feng 1 and 2 are short and medium range missile systems that are highly <u>accurate</u> and reliable under adverse weather conditions.

A. dubitable

B. punctilious

C. contestable

6. Fitted with a semi-active radar homing seeker, the Ting-Kung 1 missile is designed for medium range <u>interception</u> under an aerial saturation attack.

A. acquiescence

B. interference

C. malleability

7. The Ting-Kung 2 missile has increased range and <u>firepower</u> by adopting an active radar homing seeker.

A. appeasement

B. pacifism

C. pyrotechnics

8. Additionally, the Ray-Ting 2000, which is designed to <u>repel</u> an amphibious attack and provide superior firepower to conventional tube artillery, is equipped with a fully automatic fire control system, elevation and azimuth driven systems, as well as positional and direction determining systems.

A. repulse

B. allure

C. entice

9. Additionally, the Ray-Ting 2000, which is designed to repel an amphibious attack and provide <u>superior</u> firepower to conventional tube artillery, is equipped with a fully automatic fire control system, elevation and azimuth driven systems, as well as positional and direction determining systems.

A. slipshod

B. amateurish

C. transcendent

10. This system is considered one of the most <u>powerful</u> artillery multiple launch rocket systems worldwide.

A. enervate

B. stalwart

C. debilitated

11. To create a balance in the arms buildup involving Taiwan and China, local arms manufacturers such as the Chung Shan Institute of Science and Technology must implement appropriate management strategies to increase their ability to supply domestic defense needs, thus <u>alleviating</u> over dependence on imported weapons.

A. diminishing

B. exacerbating

C. aggravating

Unit Seven

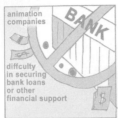

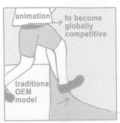

Introducing an Industry

工業介紹

1. Briefly highlight the general characteristics of this industry in Taiwan.
   此種產業在台灣的一般特色
2. Point out the difficulties encountered in industrial development.
   產業所面臨的困境
3. Describe current activities of the particular industrial sector.
   簡述一或二個目前活動的重點
4. Elaborate on the available technologies that are employed by the industry.
   產業採用的科技
5. Discuss the related research and development facilities in Taiwan, and how they assist/collaborate with that particular industry.
   台灣相關的研究發展設備

Vocabulary and related expressions

animation sector	動畫部門
digitization	數位化
original equipment manufacturers (OEMs)	原創設備製造業者
global marketing expertise	全球性市場專門技術
a more solid economic foundation	更堅固的經濟基礎
an influx	湧進／匯集
enhancing process efficiency	強化的過程效能
aging society	老化的社會
retirement housing complexes	退休住宅社區
tenants	房客
governmental subsidized	政府的補助
in conjunction with	和……連接
recuperative facilities	恢復中的設備
enormous potential	龐大的潛能
continued economic growth	持續的經濟成長
remain abreast of	和……保持並列
environmental sustainability	環境永續性
numerous incentives	多數的動機
fiercely competitive	殘酷地競爭
long-distance transportation	長距離的運輸
tremendous capital requirements	大量的資金需求
cultivate local medical talent	培育當地的醫學天才
privileged gentry class	享有特權的上流社會人士階級
severe fiscal imbalance	嚴重的財政不平衡

Situation 1

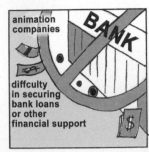  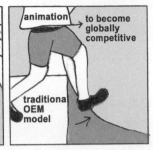

Situation 2

 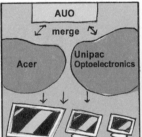 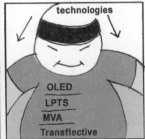

Situation 3

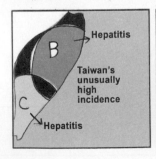 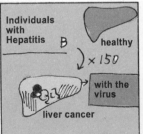 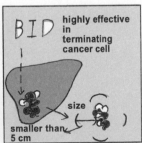

A Write down the key points of the situations on the preceding page, while the instructor reads aloud the script from the Answer Key. Alternatively, students can listen to the script online at www.chineseowl.idv. tw

Situation 1

_____

_____

Situation 2

_____

_____

Situation 3

_____

_____

B Oral practice I
Based on the three situations in this unit, write three questions beginning with **Why**, and answer them. The questions do not need to come directly from these situations.

Examples

Why are Taiwanese animation firms limited in their ability to build brand recognition?

Because overseas businesses continue to view them as original equipment manufacturers (OEMs)

Why do Taiwanese animation companies have difficulty in securing bank loans or other financial support as company startups?

Largely owing to their relative immaturity

1. _____

_____

2. _____

_____

3. _____

_____

C Based on the three situations in this unit, write three questions beginning with *What*, and answer them. The questions do not need to come directly from these situations.

Examples

What must the government do to look after the welfare of the elderly?

Provide affordable, basic housing that suits the needs of this growth sector

What did Taiwan officially become in 1997 according to the definition of the United Nations?

An aging society

1. _____

_____

2. _____

_____

3. _____

_____

D Based on the three situations in this unit, write three questions beginning with *Which*, and answer them. The questions do not need to come directly from these situations.

Examples

Which trend creates enormous potential for further development of 3C and synthetic fiber products?

The one using nano-materials in fiber production

Which applications for locally produced 3C and synthetic fiber products include inorganic nano-recorder media, nano-interface handle for a separate membrane in batteries and electron passive elements?

Nano-material ones

1. _____

_____

2. _____

_____

3. _____

_____

E Write questions that match the answers provided.

1. _____

   _____

   _____

   Inorganic nano-recorder media, nano-interface handle for a separate membrane
   in batteries and electron passive elements

2. _____

   _____

   _____

   Numerous incentives for local companies to develop their research capabilities

3. _____

   _____

   _____

   The ability to adopt advanced production technologies such as nano-materials

F Listening Comprehension I

Situation 1

1. Why are Taiwanese animation firms limited in their ability to build brand recognition?

   A. Labor costs in China and India are significantly lower than in Taiwan.

   B. Because overseas businesses continue to view them as original equipment manufacturers (OEMs)

   C. Taiwanese animation companies lack animation designers with global marketing expertise.

2. What has transformed the Taiwanese animation sector and increased productivity and innovativeness?

   A. digitization, or placing animated objects in a digital context

   B. a loan fund to develop digital content

   C. efforts to move beyond the traditional OEM model and develop brand recognition

3. Why do Taiwanese animation companies have difficulty in securing bank loans or other financial support as company startups?

   A. largely owing to the lack of animation designers with global marketing expertise

   B. largely owing to the inability to move beyond the traditional OEM model and develop brand recognition

   C. largely owing to their relative immaturity

4. Why did the Ministry of Economic Affairs recently initiate a loan fund to develop digital content?

   A. to enhance business-related research and development tasks, such as enhancing process efficiency

275

B. to benefit the animation sector

C. to apply increasingly novel 3D techniques in cartoons, animation videos and Internet-based animation and animation films

5. Which organizations in Taiwan focus on developing 3D techniques and animation designs?

A. the Ministry of Economic Affairs

B. academic research centers

C. R&D departments in animation firms

Situation 2

1. Why does the government need to provide affordable, basic housing that suits the needs of this growth sector?

A. to organize social activities, health management, hotel type services, and specialist service staff

B. to offer a diverse array of services to high income elderly individuals

C. to look after the welfare of the elderly

2. Who are long-term residential care facilities available for?

A. high income elderly individuals

B. chronically ill elderly who are largely bed ridden

C. low income elderly individuals

3. How long are luxury accommodations for elderly tenants leased to?

A. from one to three years

B. or more than twenty years

C. from three to twenty years

4. What are examples of long-term residential care facilities available for chronically ill elderly?

A. residential facilities for retired servicemen that combine housing with recuperative facilities

B. residential facilities for high income elderly individuals

C. retirement housing complexes with luxury accommodations

5. How are government-registered residential facilities run?

A. in conjunction with general nursing staff

B. in conjunction with local construction enterprises

C. in conjunction with private recuperative centers

Situation 3

1. What does continued economic development in Taiwan depend on?

A. the ability to constantly adopt advanced manufacturing practices such as nano-materials technology

B. the ability to further understand the properties of nano-materials and their implications for product commercialization

C. the ability to adopt advanced production technologies such as nano-materials

2. What is important to Taiwan's continued economic growth?

A. the ability to constantly adopt advanced manufacturing practices such as nano-materials technology

B. the ability to further understand the properties of nano-materials and their implications for product commercialization

C. the ability to adopt advanced production technologies such as nano-materials

3. What has the Taiwanese government offered numerous incentives for?

A. reducing the time between product development and commercialization

B. local companies to develop their research capabilities

C. using energy efficiently for manufacturing purposes

4. How much has the Taiwanese economy matured?

   A. to the extent where labor and capital resources are already efficiently allocated for manufacturing

   B. to the extent where the time between product development and commercialization is reduced

   C. to the extent where the properties of nano-materials are manipulated to meet industrial specifications

5. What is Taiwan a key global manufacturer of?

   A. nanotechnology products

   B. energy efficient materials

   C. 3C and synthetic fiber products

G Reading Comprehension I
Pick the work or expression whose meaning is closest to the meaning of the underlined word or expression in the following passages.

Situation 1

1. The Taiwanese animation sector is <u>involved in</u> animation films, cartoons, animation videos and Internet-based animation.

   A. desisting from

   B. abstaining from

   C. committed to

2. Digitization, or placing animated objects in a digital context, has transformed the sector and increased productivity and <u>innovativeness</u>.

   A. complacency

   B. tranquility

   C. creativity

3. Still, the animation sector faces many <u>obstacles</u>.

   A. impediments

   B. access

   C. ingress

4. Overseas businesses continue to view Taiwanese animation firms as original equipment manufacturers (OEMs), limiting their <u>ability</u> to build brand recognition.

   A. competence

   B. incapacitation

   C. impuissance

5. Additionally, <u>labor</u> costs in China and India are significantly lower than in Taiwan.

   A. regulation

   B. administration

   C. proletariat

6. Moreover, Taiwanese animation companies lack animation designers with global marketing <u>expertise</u>.

   A. callowness

   B. adeptness

   C. incompetence

7. Taiwanese animation companies also have difficulty in <u>securing</u> bank loans or other financial support as company startups, largely owing to their relative immaturity.

   A. restraining

   B. procuring

   C. detaining

8. Despite these obstacles, the local animation sector is increasingly <u>pivotal</u> in the economic development of Taiwan, and several encouraging developments are underway.

   A. paltry

   B. frivolous

   C. exigent

9. For example, the Ministry of Economic Affairs recently <u>initiated</u> a loan fund to develop digital content, which will benefit the animation sector.

   A. commenced

   B. retrenched

   C. truncated

10. In addition to providing the animation sector a more <u>solid</u> economic foundation, this project will also encourage an influx of skilled technical and management personnel.

A. piddling

B. ethereal

C. stout

11. Additionally, the Taiwanese animation sector has adopted increasingly <u>novel</u> 3D techniques in cartoons, animation videos and Internet-based animation and animation films, for example Toy Story and SHREK.

A. obsolete

B. neoteric

C. archaic

12. Organizations in Taiwan committed to developing animation-related techniques can be <u>categorized</u> as either academic research centers or R&D departments in animation firms, with the former focused on developing 3D techniques and animation designs and the latter focused on business-related research and development tasks, such as enhancing process efficiency.

A. frenzied

B. specified

C. disorganized

13. If the Taiwanese animation sector is to become globally <u>competitive</u>, it must move beyond the traditional OEM model and develop brand recognition.

A. stagnant

B. waning

C. emulous

Situation 2

1. Taiwan <u>officially</u> became an aging society in 1997 according to the definition of the United Nations.

   A. illicitly

   B. legitimately

   C. unlawfully

2. To look after the welfare of the elderly, the government needs to provide <u>affordable</u>, basic housing that suits the needs of this growth sector.

   A. costly

   B. economical

   C. excessive

3. Local construction <u>enterprises</u> thus have begun promoting retirement housing complexes, which can be categorized into four types.

   A. medical clinics

   B. governmental organizations

   C. companies

4. First, <u>luxury</u> accommodations are available for retirees, which have restrictions on tenant age and health status.

   A. opulent

   B. meat-and-potatoes

   C. nitty-gritty

5. Leased to tenants for long periods ranging from three to twenty years, such communities are normally referred to as congregate housing, and are characterized by <u>uniform</u> housing, organized social activities, health management, hotel type services, and specialist service staff.

   A. diversified

   B. potpourri

C. homogeneous

6. Such housing arrangements offer a diverse array of services to <u>high income</u> elderly individuals.

   A. affluent

   B. impoverished

   C. destitute

7. Second, government-registered residential facilities are available for the elderly, which are subsidized by the government with elderly residents contributing the <u>remainder</u> themselves.

   A. preponderance

   B. remnant

   C. lion's share

8. Such facilities provide only <u>basic</u> daily services and so are relatively inexpensive.

   A. cut-rate

   B. exorbitant

   C. excessive

9. Third, government-registered residential facilities are run <u>in conjunction with</u> private recuperative centers.

   A. in contrast to

   B. in line for

   C. in line with

10. These facilities mainly house bed-ridden <u>elderly</u>, and are staffed by general nursing staff without specialized skills and who simply provide primary care.

    A. adolescent

    B. juvenile

    C. aged

11. Fourth, long-term residential care facilities are available for chronically ill elderly who are <u>largely</u> bed ridden.

    A. seldom

    B. to a large extent

    C. scarcely

12. These facilities <u>differ from</u> long-term care facilities in hospitals in that they aim mainly to support the daily functions of the elderly, while hospitals focus on treating chronic illnesses.

    A. correlate with

    B. correspond to

    C. contrast with

13. Notable examples of these facilities include residential facilities for retired servicemen that combine housing with <u>recuperative</u> facilities.

    A. degenerative

    B. restorative

    C. regressive

14. What all of the above facilities share in common is that they are residential nursing homes staffed with <u>professional</u> nurses.

    A. bush-league

    B. white-collar

    C. a poor excuse for

Situation 3

1. Taiwan is a <u>key</u> global manufacturer of 3C and synthetic fiber products.

    A. trifling

    B. inconsequential

    C. predominant

2. The current trend towards using nano-materials in fiber production creates enormous potential for further development of this industry.

A. minuscule

B. stupendous

C. infinitesimal

3. The numerous existing nano-material applications for <u>locally</u> produced 3C and synthetic fiber products include inorganic nano-recorder media, nano-interface handle for a separate membrane in batteries and electron passive elements.

A. globally

B. internationally

C. domestically

4. The ability to adopt advanced production technologies such as nano-materials is important to Taiwan's <u>continued</u> economic growth.

A. sustained

B. intermittent

C. sporadic

5. Taiwanese manufacturers must remain abreast of the latest applications of nano-elements and organisms, while also considering environmental <u>sustainability</u> and using energy efficiently for manufacturing purposes.

A. burden

B. costs

C. sustenance

6. Since the Taiwanese economy has <u>matured</u> to the extent where labor and capital resources are already efficiently allocated for manufacturing, continued economic development depends on the ability to constantly adopt advanced manufacturing practices such as nano-materials technology.

A. deteriorated

B. decayed

C. perfected

7. Accordingly, the Taiwanese government has offered numerous <u>incentives</u> for local companies to develop their research capabilities, and these incentives have seen Taiwan transform from an agricultural-based economy to a hi-tech one over just a few decades.

A. retributions

B. impetuses

C. hindrances

8. During the coming decade, Taiwan faces the following challenges in further developing its nano-materials technology sector: developing and synthesizing nano-materials efficiently, <u>manipulating</u> the properties of nano-materials to meet industrial specifications, and further understanding the properties of nano-materials and their implications for product commercialization.

A. acquiesce

B. concurring with

C. coercing

9. One particular concern is how to <u>reduce</u> the time between product development and commercialization.

A. abridge

B. augment

C. amass

10. Taiwan must effectively meet the above challenges to remain competitive in the <u>fiercely</u> competitive nanotechnology sector.

A. quiescent

B. placid

C. tempestuously

H　Common elements in introducing an industry工業介紹include the following elements:

1. Briefly highlight the general characteristics of this industry in Taiwan.
   此種產業在台灣的一般特色

· Taiwan's medical sector encompasses administrative organizations, hygienic safety units throughout all levels of hospital operations, pharmaceutical manufacturers, medical technology enterprises, academic institutions, information technology providers and, most importantly, patients.　More than hospitals, Taiwan's medical sector also comprises smaller clinics and other medical organizations.

· While Taiwan has focused on developing the semiconductor and OEM industries in recent decades, the question arises as from where the island's latest technological advancements will emerge. Restated, exactly where does Taiwan rank in the globalization scheme? With the rapidly elderly population worldwide, the long term health sector will undoubtedly emerge as a leading industry.

· Characterized as low polluting and capital intensive, the biotechnology industry employs highly qualified personnel from multidisciplinary fields and develops product technologies with a high economic return. Given those characteristics, R&D budget accounts for a large proportion of the company's operating budget, which is understandable if a company wishes to enhance its competitiveness.

2. Point out the difficulties encountered in industrial development.
   產業所面臨的困境

· Among the several challenges that the automotive industry in Taiwan has faced include the extended time required for automobile design, subsequently increasing new product delivery time to the market; large control that the overseas headquarters maintains in which technologies to adopt and how original models should be modified, thus limiting the design capacity of the local manufacturer; and inability of the overseas headquarters to directly promote the competitiveness of its products in the local market sector besides the quality of the car brand itself.

· The bedding sector faces several obstacles, such as the reluctance of customers to spend more than $5,000 New Taiwanese (NT) dollars on a mattress. Alternatively, local repair companies often fix spring beds or purchase old mattresses to recycle their parts. Although economically attractive, the repair and recycle processes are often environmentally unfriendly. Also, many consumers complain of discomfort after sleeping on a spring bed mattress. Therefore, many companies find it difficult to sell higher priced mattresses although superior in quality to traditionally used spring bed mattresses.

· Among the over 2000 waste treatment companies operate in Taiwan, only ten are qualified to handle medical waste from hospitals. Among those, four are not of large industrial scale and most do not have over 100 staff. Given the small scope of these businesses, services significantly vary and non-uniform treatment standards often lead to environmental contamination.

3. Describe current activities of the particular industrial sector.

    簡述一或二個目前活動的重點

- · The Taiwanese government has recently promoted a new policy entitled "Broaden the sources of National Health Insurance generated income to reduce medical expenditures" as an incentive for hospital administrators to increase productivity and efficiency, as well as assure the general public of the continued provision of high quality medical care.
- · The paper processing industry significantly contributes to domestic and overseas forestation efforts. For instance, Yuen Foong Yu Paper Corporation and the Chinese Pulp Corporation have collaborated in forestation of 2,000 hectares in Vietnam, with Yuen Chi Paper Corporation investing in 150,000 hectares of forestation in Indonesia.
- · Most Taiwanese biotechnology firms focus on developing therapeutic products, diagnostics, medicinal applications through plant extracts, agricultural production and even genetic testing.

4. Elaborate on the available technologies that are employed by the industry.

   產業採用的科技

- · To address this concern, several PACS manufacturers have recently entered this burgeoning market by introducing advanced manufacturing capabilities at a more competitive retail cost given the long-term investment involved in operating and maintaining this system.
- · In addition to enhanced training of research and managerial personnel in specific fields, local biotechnology and medical organizations are

exploring ways to integrate other innovative fields such as bioinformatics and genomics into their research scope and product line.

· In terms of the development of GPS navigational systems, domestic manufacturer capacity is nearly equivalent to that of overseas automakers although locally developed automotive security systems and wireless on-line network capabilities still lag behind overseas counterparts. Given rapid scientific and technological advances, the automotive industry will heavily emphasize wireless on-line network capabilities, GPS navigational features and foolproof security systems.

5. Discuss the related research and development facilities in Taiwan, and how they assist/collaborate with that particular industry.
台灣相關的研究發展設備

· Designed by the University of Wisconsin as the future of image-guided IMRT, tomotherapy is essential therapeutic instrumentation in treating lung cancer patients owing to its ability to allow precise image-guided intensity modulated radiotherapy and provide valuable information regarding tumor changes during radiotherapy. The ability of local manufacturers to collaborate with overseas partners will determine whether Taiwan can provide quality assurance treatment for ailing patients.

· Medical organizations in Taiwan are divided into four categories based on the nature of services offered. First, primary medical hospitals provide general medical health service and continuous medical attention in a clinical environment. Second, local hospitals offer general specialized services in outpatient, inpatient and emergency care units, with patients offered primary care. Third, besides offering services covered in the first two categories, regional hospitals provide residential training for intern physicians, thus functioning as a teaching hospital. Finally, medical centers cover a diverse array of services, including research, teaching, training and highly specialized medical practices and operations.

· Given the potential of collagen in restoring skin growth, several manufacturers have devoted considerable resources to product development, including the Taiwan Salt Industrial Corporation, Taiwan Sugar Corporation, Taiwan Fertilizer Corporation and Chang Gung Biotechnology Corporation.

In the space below, introduce an industry.

_____

_____

_____

_____

_____

_____

_____

_____

_____

Look at the following examples of introducing an industry.

Taiwan's medical sector encompasses administrative organizations, hygienic safety units throughout all levels of hospital operations, pharmaceutical manufacturers, medical technology enterprises, academic institutions, information technology providers and, most importantly, patients. More than hospitals, Taiwan's medical sector also comprises smaller clinics and other medical organizations. Hospital operations also encompass a vast array of services, including general practice, specialized treatment, treatment of the chronically ill, psychiatric consultancy, traditional Chinese medicine and orthodontist treatment. Among the clinical services offered include specialized treatment, general practice, traditional Chinese medicine, and orthodontist treatment. Other medical organizations include blood donor organizations, pathology organizations and other supporting agencies engaged in other medical treatment and health care management that do not directly treat patients. Medical organizations in Taiwan are divided into four categories based on the nature of services offered. First, primary medical hospitals provide general medical health service and continuous medical attention in a clinical environment. Second, local hospitals offer general specialized services in outpatient, inpatient and emergency care units, with patients offered primary care. Third, besides offering services covered in the first two categories, regional hospitals provide residential training for intern physicians, thus functioning as a teaching hospital. Finally, medical centers cover a diverse array of services, including research, teaching, training and highly specialized medical practices and operations. As the primary teaching hospital of a medical college, a medical center is responsible for teaching clinical medical science. Several unique features of Taiwan's current circumstances have significantly impacted the island's medical sector, including the dramatic rise in chronic illnesses and a rapidly aging population. Additionally, elevated living standards have increased awareness among the general public of the importance of medical treatment and long-term care for aging residents. Moreover, abuse of medical resources in the National Health Insurance scheme has led to drastic reform measures.

As automobiles have been designed and manufactured for more than a century, continuous development of this sector reflects consumer preferences with respect to style and available functions. Among the several challenges that the Taiwanese automotive industry has faced include the extended time required for automobile design, subsequently increasing new product delivery time to the market; large control that the overseas headquarters maintains in which technologies to adopt and how original models should be modified, thus limiting the design capacity of the local manufacturer and making it impossible for the overseas headquarters to directly promote the competitiveness of its products in the local market sector. Despite these challenges, the local automotive sector has witnessed several innovative changes recently. For instance, the installation of a multimedia entertainment system for rear seat passengers is a highly promising area for further technological development and heightened commercial interest. While adhering to governmental safety regulations, design of this entertainment system does not distract the driver from concentrating on the road ahead. According to projections from the Industrial Economic and Knowledge (IEK) Center of Industrial Technology Research Institute, North American trends in the rear seat entertainment system will change over the next two years, advancing to levels that will incorporate GPS navigational and telecommunications features, as well as an informatics system. In terms of the development of GPS navigational systems, domestic manufacturer capacity is nearly equivalent to that of overseas automakers although locally developed automotive security systems and wireless on-line network capabilities still lag behind overseas counterparts. Given rapid scientific and technological advances, the automotive industry will heavily emphasize wireless on-line network capabilities, GPS navigational features and foolproof security systems.

The biotechnology industry is globally recognized as the technological pacesetter for the new century. As the industry heavily emphasizes establishing innovative commercial ventures that match biotechnology developments, nearly all countries prioritize this industry in their national development strategies. Characterized as low polluting and capital intensive, this industry employs highly qualified personnel from multidisciplinary fields and develops product technologies with a high economic return. Given those characteristics, R&D budget accounts for a large proportion of the company's operating budget, which is understandable if a company wishes to enhance its competitiveness. Among numerous new knowledge economy-oriented ventures, biotechnology plays a leading role in scientific and technological related research. For instance, after Taiwan Sugar Corporation（TSC）commercialized its placenta extract product, many local enterprises followed suit in entering the bio-cosmetics market, including Taiwan Salt Industrial Corporation and the Formosa Plastics Group. In this sector, nanometer-based technology is increasingly replacing conventionally adopted technical capabilities in product design. As the local bio-cosmetics sector continues to expand, many industries have expressed interest in expanding their product lines to this area. Whether TSC maintains its leading position in domestic bio-cosmetics production depends on its ability to continuously generate new products and more fully understand consumer trends. Still, Taiwanese bio-cosmetics manufacturers face several obstacles, including a small domestic market, lack of investment from the government and other sources, reliance on imported technologies and lack of multidisciplinary personnel committed to this area of product technology research. Ultimately, R&D activities will determine whether the domestic bio-cosmetics sector can compete in the global market. As in other sectors, cosmetic manufacturers devote considerable resources to developing new products. In particular, an increasing number of companies strive to incorporate consumer preferences into their product development efforts, which requires constantly surveying customers' attitudes and remaining abreast of the latest trends. Given the unique features of product development in Taiwan's cosmetic sector, manufacturing technologies adopted have matured, resulting in enhanced product quality. Cosmetic manufacturers especially focus on female consumers, who form a vital niche in the market, given the strong desire to appear as youthful as possible. Manufacturers increasingly realize that remaining competitive depends on the ability to understand the unique features of the island's cosmetic sector. For instance, given the potential of collagen in restoring skin growth, several manufacturers have devoted considerable resources to product development, including the Taiwan Salt Industrial Corporation, Taiwan Sugar Corporation, Taiwan Fertilizer Corporation and Chang Gung Biotechnology Corporation. To launch new product lines efficiently, many companies adopt a multidisciplinary approach by integrating the efforts of various departments, each having its own expertise and contribution towards product research. Synergy among these departments resolves obstacles to product innovation and communication channels, thus enhancing the efficiency of product development and a timely delivery to the market. Moreover, the biotech industry plays a prominent role in the cosmetic manufacturing sector, with its ability to develop innovative products largely determining corporate survival in this intensely competitive environment.

Situation 4

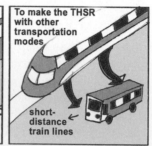

Situation 5

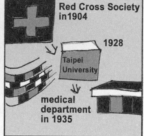
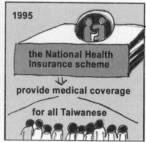

Situation 6

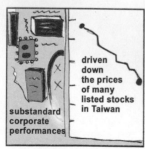

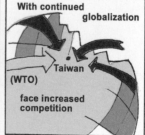

I Write down the key points of the situations on the preceding page, while the instructor reads aloud the script from the Answer Key. Alternatively, students can listen to the script online at www.chineseowl.idv. tw

Situation 4

_____

_____

_____

Situation 5

_____

_____

_____

Situation 6

_____

_____

J   Oral practice II
Based on the three situations in this unit, write three questions beginning with **What**, and answer them. The questions do not need to come directly from these situations.

Examples

What organization oversees conventional rail transportation in Taiwan?

The Taiwan Rail Administration (TRA)

What have not justified the tremendous capital requirements of railway construction?

Recent passenger numbers

1. _____

_____

2. _____

_____

3. _____

_____

K Based on the three situations in this unit, write three questions beginning with **When**, and answer them. The questions do not need to come directly from these situations.

Examples

When did Japan seek to cultivate local medical talent by discouraging its own physicians from practicing in Taiwan?

During its occupation of Taiwan

When did the Japanese establish a medical training institute at a teaching hospital facility in Taipei?

In 1897

1. _____

_____

2. _____

_____

3. _____

_____

L Based on the three situations in this unit, write three questions beginning with **Why**, and answer them. The questions do not need to come directly from these situations.

Examples

Why has the Taiwanese government has initiated comprehensive reforms focused on market liberalization and institutional mergers?

To strengthen the financial sector

Why have merged entities suffered an impaired corporate performance?

Owing to defaulted loans and debts acquired during their mergers

1. _____

   _____

2. _____

   _____

3. _____

   _____

M Write questions that match the answers provided.

_____

_____

By initiating comprehensive reforms focused on market liberalization and institutional mergers

_____

_____

Strengthen poorly performing banks or credit cooperatives by merging them with more healthy financial institutions

_____

_____

As overseas firms gain a foothold in the domestic market

N Listening Comprehension II

Situation 4

1. How long does the Taipei-Kaohsiung Tzu-Chiang express train ride last?

    A. 3 hours 59 minutes one way

    B. 3 hours 59 minutes round trip

    C. 4 hours 59 minutes one way

2. What does not justify the tremendous capital requirements of railway construction?

    A. expensive off-peak ticket prices

    B. lack of governmental funding

    C. recent passenger numbers

3. Why have individuals been encouraged to drive cars or take passenger buses instead of riding the train?

    A. custom-designed train tourism packages

    B. expanded train cargo capacity

    C. completion of the Chung-shan and Formosa freeway projects

4. What does TRA need to do in order to make the THSR competitive with other transportation modes?

    A. encourage automotive companies to offer more inexpensive cars

    B. provide additional short-distance train lines and promote custom-designed train tourism packages

    C. offer luxurious seating and inexpensive off-peak ticket prices

5. Why should the Taiwan Department of Transportation integrate the efforts of TRA and the THSR?

    A. to provide a comprehensive, comfortable and convenient passenger service

    B. to purchase the push-pull Tzu-Chiang express train and, most significantly, the south-link line and the round island railway network

C. to construct a rail line from Keelung to Taipei

Situation 5

1. How many regional hospitals operate in Taiwan?

    A. 16

    B. 70

    C. 375

2. Why was a large island-wide protest march of medical professionals held this year?

    A. a severe fiscal imbalance, causing near bankruptcy and other problems for medical facilities that rely on NHI subsidies

    B. the emergence of unlicensed physicians and poor quality medicines

    C. inflationary prices and scarcity of medicine

3. How long have medical practices and standards in Taiwan gradually improved?

    A. since 1928

    B. since 1965

    C. since 1995

4. What was established in 1904?

    A. the Japanese Red Cross Society in Taiwan

    B. a medical department at National Taiwan University

    C. a medical training institute in Taiwan

5. Why did Japan discourage its own physicians from practicing in Taiwan during its occupation of Taiwan?

    A. to cater mainly to Japanese governmental officials or the privileged gentry class in Taipei

    B. to provide medical coverage for all Taiwanese residents

    C. to cultivate local medical talent

Situation 6

1. Why have merged entities suffered from an impaired corporate performance?

    A. owing to continued globalization, most notably in the form of WTO entry

    B. owing to the foothold that overseas firms have gained in the domestic market

    C. owing to defaulted loans and debts acquired during their mergers

2. How were merger policies intended to strengthen poorly performing banks or credit cooperatives?

    A. by merging them with more healthy financial institutions

    B. by increasing banking service quality and improving the competitiveness of domestic institutions

    C. by increasing loan default rates

3. What comprehensive reforms in the financial sector has the Taiwanese government initiated?

    A. those focusing on listed stocks in Taiwan

    B. those focusing on Taiwan's entry into the WTO

    C. those focusing on market liberalization and institutional mergers

4. Why have large numbers of banks emerged?

    A. the increased competitiveness of domestic institutions

    B. market liberalization policies

    C. well-intentioned governmental policies

5. Why have the prices of many listed stocks in Taiwan been driven down?

    A. substandard corporate performances

    B. increasing loan default rates

    C. market liberalization and institutional merger policies

O Reading Comprehension II
Pick the work or expression whose meaning is closest to the meaning of the underlined word or expression in the following passages.

Situation 4

1. The Taiwan Rail Administration (TRA) <u>oversees</u> conventional rail transportation in Taiwan.

   A. subsidizes

   B. orchestrates

   C. transfers

2. Since Liu Ming-Chuan directed the construction of the first rail line from Keelung to Taipei in 1886, rail has been <u>important</u> to long-distance transportation in Taiwan.

   A. insignificant

   B. irrelevant

   C. critical

3. In response to rapid population and economic growth, TRA completed the north-link line, the double link of the western line, the electrified project of the western line, the <u>purchase</u> of the push-pull Tzu-Chiang express train and, most significantly, the south-link line and the round island railway network.

   A. procurement

   B. investment

   C. lease

4. Recent passenger numbers have not <u>justified</u> the tremendous capital requirements of railway construction.

   A. extirpated

B. expunged

C. warranted

5. Notably, completion of the Chung-shan and Formosa freeway projects <u>encouraged</u> individuals to drive cars or take passenger buses instead of riding the train, and many bus companies offer an attractive combination of luxurious seating and inexpensive off-peak ticket prices.

A. enticing

B. repulsive

C. nauseating

6. To encourage passengers to return to railway transportation, TRA has promoted the Taipei-Kaohsiung Tzu-Chiang express train (3 hours 59 minutes one way) and the Taipei-Hualien Chu-Kuang group express train. The Taiwan High Speed Railway (THSR) has been the most <u>daunting</u> undertaking of TRA to date.

A. diffident

B. timid

C. adventurous

7. To realize the THSR, TRA is <u>directing</u> several construction projects, including commuter stations in San-Keng, Tai-Yuan, Da-Ching and Da-Chiao, as well as a metro hub linking Taipei and Hsinchu.

A. monitoring

B. coordinating

C. identifying

8. To make the THSR <u>competitive</u> with other transportation modes, TRA needs to provide additional short-distance train lines, promote custom-designed train tourism packages, provide shuttle services linking THSR stations with downtown areas and expand train cargo capacity.

A. lagging behind

B. emulative

C. waning

9. To make the THSR competitive with other transportation modes, TRA needs to provide additional short-distance train lines, promote custom-designed train tourism packages, provide shuttle services linking THSR stations with downtown areas and expand train cargo <u>capacity</u>.

A. improvements

B. quality

C. space

10. Moreover, the Taiwan Department of Transportation should integrate the efforts of TRA and the THSR to provide a comprehensive, <u>comfortable</u> and convenient passenger service.

A. tranquil

B. distressing

C. nerve-racking

Situation 5

1. During its occupation of Taiwan, Japan sought to <u>cultivate</u> local medical talent by discouraging its own physicians from practicing in Taiwan.

A. discourage

B. nurture

C. debilitate

2. Moreover, the Japanese <u>established</u> a medical training institute in 1897 at a teaching hospital facility in Taipei, which in 1899 became a medical school with its own specialized departments of medical science.

A. contrived

B. eradicated

C. effaced

3. However, this facility catered mainly to Japanese governmental officials or the privileged <u>gentry class</u> in Taipei rather than catering to the general public.

A. aristocracy

B. proletariat

C. the silent majority

4. Later, teaching hospitals were established, which <u>aimed</u> to serve local community needs. Eventually, the Japanese Red Cross Society in Taiwan was established in 1904.

A. aggravated

B. abdicated

C. attempted

5. Moreover, the Taipei University of the Japanese Prefecture <u>set up</u> a medical department in 1935, some years after the university's original establishment in 1928.

A. transformed into

B. transferred to

C. constructed

6. After Chiang-Kai Shek and his Kuomingtang Government arrived in Taiwan in the <u>wake</u> of the civil war, problems such as inflationary prices and scarcity of medicine led to the emergence of unlicensed physicians and poor quality medicines.

A. aftermath

B. beginning

C. advent

7. However, since 1965, medical practices and standards in Taiwan have gradually improved to reach the current situation where Taiwan provides <u>state-of-the-art</u> medical services.

   A. archaic

   B. antiquated

   C. progressive

8. Taiwanese hospitals are <u>categorized</u> either as medical centers (17), regional hospitals (70), community hospitals (375) or clinics (data unavailable).

   A. regulated

   B. classified

   C. outlined

9. Since its establishment in 1995, the National Health Insurance scheme has sought to provide medical coverage for all Taiwanese <u>residents</u> under the auspices of the National Health Insurance Bureau.

   A. finances

   B. direction

   C. classification

10. However, the National Health Insurance scheme suffers from a <u>severe</u> fiscal imbalance, causing near bankruptcy and other problems for medical facilities that rely on NHI subsidies.

   A. implacable

   B. facile

   C. adroit

11. Consequently, a <u>large</u> island-wide protest march of medical professionals was held this year.

   A. dwarfed

   B. stunted

   C. prodigious

Situation 6

1. Substandard corporate performances have driven down the prices of many listed stocks in Taiwan, <u>straining</u> the Taiwanese financial sector by increasing loan default rates.

    A. abating

    B. distending

    C. slackening

2. To strengthen the financial sector, the Taiwanese government has initiated <u>comprehensive</u> reforms focused on market liberalization and institutional mergers.

    A. antithetical

    B. incompatible

    C. capacious

3. However, these well-<u>intentioned</u> policies have incurred further financial problems.

    A. meaning

    B. designed

    C. structured

4. For example, market liberalization policies have led to the emergence of large numbers of banks, <u>saturating</u> the market to the point where banking institutions suffer low profits and have difficulty differentiating themselves from competitors.

    A. glancing

    B. scanning

    C. permeating

5. Additionally, institutional mergers failed to <u>resolve</u> basic financial problems.

    A. abate

    B. amplify

    C. exacerbate

6. The merger policies were intended to <u>strengthen</u> poorly performing banks or credit cooperatives by merging them with more healthy financial institutions.

A. enervate

B. invigorate

C. deplete

7. However, merged entities have suffered an <u>impaired</u> corporate performance owing to defaulted loans and debts acquired during their mergers.

A. tainted

B. healthy

C. vigorous

8. With <u>continued</u> globalization, most notably in the form of WTO entry, the financial sector will undoubtedly face increased competition as overseas firms gain a foothold in the domestic market.

A. abating

B. receding

C. sustained

9. With continued globalization, most notably in the form of WTO entry, the financial sector will undoubtedly <u>face</u> increased competition as overseas firms gain a foothold in the domestic market.

A. disavow

B. encounter

C. refute

10. Furthermore, given the financial problems associated with market liberalization and institutional merger <u>policies</u>, the Taiwanese government must closely re-examine the problems and implement responses to increase banking service quality and improve the competitiveness of domestic institutions.

A. recruiting procedures

B. procurement strategies

C. measures

Answer Key

解答

Answer Key
Forecasting Market Trends
預測市場趨勢

A

Situation 1

E-learning in Taiwan has recently emerged as a highly promising learning medium for enhancing traditional classroom instruction. For example, e-learning eliminates time constraints and space limitations faced by classroom instruction. Another strength of e-learning is its ability to construct personal learning environments. Internet-based English learning websites are a notable example. Such websites commonly adopt the sharable content object reference model (SCORM), which emphasizes reusability, accessibility, durability and interoperability. Successful adoption of the SCORM standard has significantly reduced the temporal and spatial constraints faced by e-learners. Unsurprisingly, recent statistics indicate strong growth in e-learning. For instance, in 2003, the e-learning market in the United States generated revenues of US$ 400,000,000, and the compound annual growth rate for e-learning revenues is predicted to reach 20.7% from 2002 to 2007. In sum, e-learning represents a highly promising area for digital content applications. Commercial strategies involving Internet-based English learning websites thus are receiving increased attention. Successful commercialization requires prioritizing market orientation, making teaching design and pedagogical content a primary concern. Interactive websites enable learners to enhance their competitiveness in school and at work. Successful learning websites adopt the latest information technologies while integrating the expertise of multidisciplinary professionals. For instance, several Internet-based English learning websites employ professionals in art design, information technology, marketing and curriculum design. Since potential revenues increase with market size, Internet-based English learning websites should compete not only for the Taiwan market, but also for the Greater China market. E-learning approaches and related expertise can also be applied for developing asynchronous learning and vocational training websites.

Situation 2

Since the establishment of the National Health Insurance scheme in 1995 and the National Health Information Network (HIN), medical-oriented information technology has been extensively adopted in clinical practice, as evidenced by widespread information system outsourcing. An increasing number of hospitals and medical institutions purchase commercially produced medical information systems or related components. There is increasing reliance on information technology firms to design medical information systems and software. However, creating medical software involves several complex issues such as effectively integrating information. While information technology firms can develop software to match individual hospital needs, integrating future applications into existing systems can be problematic. Computerization of hospital operations increased from 28% in 1994 to 57% in 1996. Available medical information systems include medical management systems, medical Intranet systems, Internet-based medical systems and electronic charts. According to Information Security Technical Report (Vol.1, No3), the Outsourcing Institute forecasts annual expenditures on outsourcing by organizations in the United States at $100 billion dollars. Information technology (IT) outsourcing accounted for 40% of this total, or $40 billion dollars. The US information industry outsourced over $US 1,755,000,000 to the medical services sector in 2001, increasing to $US 2,809,000,000 in 2005. As evidenced by an average annual compound growth rate of 12.5%, information systems outsourcing in the medical services sector in the United States is clearly growing. The largest area for outsourcing continues to be information technology. The items most likely to be outsourced are hardware maintenance, training, applications development, re-engineering and mainframe data centers. Therefore, the main areas in terms of marketing and direct sales to companies are servers, software applications, maintenance, networks, desktop systems and end-user support items.

315

Situation 3

The biotech sector has emerged as one of the main areas of industrial growth in the new century owing to its profound impact on human quality of life. As part of its efforts to prioritize this area in national development strategies, the Taiwanese government has offered numerous incentives for investment in and development of this industry, making the industry a potential engine of sustained economic growth. In the biotech sector, examination reagents, medical supplies and medicaments are all key areas for development. Additionally, the SARS crisis in 2003 heightened concerns over Taiwan's inability to effectively prevent epidemics and treat infected patients. Efforts in drug and vaccine development following the SARS epidemic demonstrated the increased attention being paid to biotechnology. Besides raising awareness of the need to effectively prevent diseases or develop a successful vaccine, the SARS crisis emphasized the need to promote immunity in the community. While biotechnology efforts in Taiwan are focused on the pharmaceutical industry, the local market scale is small owing to limited success in penetrating the household market, as evidenced by the lack of research and development capabilities, incomplete clinical research and product trials, as well as the daunting regulations governing the use of specific drugs. To foster its competitiveness, the biotechnology sector should strengthen its research capabilities, develop patented technologies and attract biotechnology professionals with expertise in multidisciplinary fields.

B

What do recent statistics indicate?
Strong growth in e-learning

What does successful commercialization require?
Prioritizing market orientation, making teaching design and pedagogical content a

primary concern

What can e-learning approaches and related expertise also be applied for developing?

Asynchronous learning and vocational training websites

C

When did computerization of hospital operations increased from 28% to 57%?

Between 1994 to 1996

When did the US information industry outsource over $US 1,755,000,000 to the medical services sector?

In 2001

When did the US information industry outsource $US 2,809,000,000 to the medical services sector?

In 2005

D

Why were concerns heightened over Taiwan's inability to effectively prevent epidemics and treat infected patients?

Because of the SARS crisis in 2003

Why was increased attention paid to biotechnology following the SARS epidemic?

Because of efforts in drug and vaccine development

Why is the local market scale for biotechnology small?

Owing to limited success in penetrating the household market, as evidenced by the lack of research and development capabilities, incomplete clinical research and product trials, as well as the daunting regulations governing the use of specific drugs

E

What did the SARS crisis in 2003 heighten concerns over?
Taiwan's inability to effectively prevent epidemics and treat infected patients

What should the biotechnology sector do to foster its competitiveness?
Strengthen its research capabilities, develop patented technologies and attract biotechnology professionals with expertise in multidisciplinary fields

Why has the biotech sector emerged as one of the main areas of industrial growth in the new century?
Owing to its profound impact on human quality of life

F

Situation 1

1. B 2. A 3. C 4. C 5. A

Situation 2

1. C 2. A 3. C 4. A 5. C

Situation 3

1. C 2. B 3. C 4. A 5. A

G

Situation 1

1. C 2. A 3. A 4. C 5. A 6. C 7. B 8. A 9. C 10. B 11. B 12. C
13. B 14. A 15. C 16. B 17. B 18. A

Situation 2

1. B 2. B 3. A 4. C 5. A 6. B 7. A 8. C 9. A 10. A 11. C 12. B
13. B 14. A

Situation 3

1. C 2. A 3. A 4. C 5. B 6. A 7. C 8. B 9. A 10. C

Situation 4

Established in 2000, the Taipei Smart Card Corporation (TSCC) was set up by cooperation among the Taipei City Government, the Taipei Rapid Transit Corporation, 13 private bus companies, TAIPEIBANK, Mitac Inc., Taishin International Bank, and various others. TSCC has implemented a contact free smart-card ticketing system for buses, the metro and public off-road car parks in Taipei. The establishment of TSCC marks a milestone for intelligent transport systems in Taiwan. In the future, TSCC will expand its services to other areas. Unlike the conventional magnetic card, EASYCARD is the first IC card for use in mass transportation in Taiwan. EASYCARD transactions are executed wirelessly through memory IC chips and induction circuits implanted in the card. With features of large capacity, durability, speed, accuracy and security, EASYCARD enables prompt transactions and has a long lifetime. EASYCARD frees passengers from carrying coins or tokens, or making repetitive ticket purchases. Additionally, transfer rides do not require advance ticket validation, thus enhancing user convenience. EASYCARD conveniently combines payment for several transport modes into one ticket. Finally, EASYCARD represents reduced traffic congestion, as public transport utilization rates increase in response to improved service quality.

Situation 5

Market demand for chemotherapy medicine has expanded owing to the increasing incidence of cancer, accelerating research efforts to develop more effective treatments. Besides chemotherapy for treating and preventing cancer, many food products with anti-cancer claims have recently been marketed. Public fear of cancer has led to the popularization of preventive measures. For example, individuals who wish to prevent cancer are encouraged to avoid excessive drinking and smoking, get sufficient sleep, exercise, eat a nutritionally balanced diet and avoid greasy food. Additionally, several biotech firms are developing foods with anti-cancer properties. Successful product commercialization will undoubtedly yield numerous benefits. Cancer treatments include a) physical treatment, through organ removal and subsequent radiotherapy b) chemotherapy, which normally occurs after physical treatment and aims to remove all remaining cancer cells. Regarding trends in the development of chemotherapy medicine, more effective chemotherapy drugs developed in the future may be able to prevent metastasis in cancer cells. Such drugs would greatly reduce patient discomfort and the adverse effects of physical treatment during cancer therapy.

Situation 6

Continuous technological advances have popularized the installation of digitalized consumer products in automobiles in Taiwan: from telematics to rear seat entertainment systems. Automobiles are gradually becoming communications centers, with future automobile design anticipated to integrate media entertainment with the vehicle computer system. Rear seat multimedia entertainment systems are a highly promising area for further technological development and are attracting increased commercial interest. The MARCH car brand was launched in Taiwan in 1993, largely owing to the technological limitations of its Japanese parent factory in manufacturing mini-sized cars. Another local car brand, Matiz, from the Formosa Automobile Corporation, focuses on cars with less than 1000CC horsepower, with a diverse array of color combinations. Meanwhile, the Solio car brand from Suzuki is intended to appeal to younger car drivers with its ethnic appeal and numerous exterior accessories, such as fins. As for retail sales, June is generally a slow month, with the most profitable period running from September through December or January. The average car owner purchases a new model every ten years. According to January 2005 statistics on automobile sales from the Taiwan Transportation Vehicle Manufacturers Association, the leading car brands in Taiwan in terms of production and sales are Suzuki (45%), Nissan (26%), Formosa (15%) and Hyundai (13%). Japanese brands thus are highly successful in Taiwan, with their emphasis on catering to individual preferences and their recreational appeal. Restated, an appealing car exterior and large variety of interior accessories with driving and entertainment functions are the foundations of the commercial success of these Japanese car brands.

J

How was the Taipei Smart Card Corporation (TSCC) established in 2000?
By cooperation among the Taipei City Government, the Taipei Rapid Transit Corporation, 13 private bus companies, TAIPEIBANK, Mitac Inc., Taishin International Bank, and various others

How is the establishment of TSCC important?
It marks a milestone for intelligent transport systems in Taiwan.

How does EASYCARD enables prompt transactions?
With features of large capacity, durability, speed, accuracy and security,

K

What has expanded owing to the increasing incidence of cancer?
The market demand for chemotherapy medicine

What has led to the popularization of preventive measures?
Public fear of cancer

What are individuals who wish to prevent cancer encouraged to do?
Avoid excessive drinking and smoking, get sufficient sleep, exercise, eat a nutritionally balanced diet and avoid greasy food

What will undoubtedly yield numerous benefits?
Successful product commercialization

What may be able to prevent metastasis in cancer cells?
More effective chemotherapy drugs developed in the future

L

Why are rear seat multimedia entertainment systems a highly promising area?
For further technological development and increased commercial interest

Why was the MARCH car brand launched in Taiwan in 1993?
Largely owing to the technological limitations of its Japanese parent factory in manufacturing mini-sized cars

Why is the Solio car brand from Suzuki intended to appeal to younger car drivers?
Because of its ethnic appeal and numerous exterior accessories, such as fins

M

What has popularized the installation of digitalized consumer products in automobiles in Taiwan?

Continuous technological advances

Why are Japanese brands highly successful in Taiwan?
Because of their emphasis on catering to individual preferences and their recreational appeal

When was the MARCH car brand launched in Taiwan?
In 1993

N
Situation 4
1. C 2. A 3. C 4. A 5. C

Situation 5
1. C 2. B 3. A 4. B 5. A

Situation 6
1. C 2. B 3. B 4. C 5. A 6. C

O
Situation 4
1. B 2. A 3. A 4. C 5. B 6. B 7. A 8. B 9. B 10. C

Situation 5
1. C 2. B 3. A 4. C 5. A 6. A 7. C 8. A 9. A 10. C

Situation 6
1. B 2. B 3. A 4. C 5. B 6. A 7. C 8. C 9. A 10. B 11. A

Answer Key
Describing Product or Service Development
產品或服務研發

A

Situation 1

With its strong emphasis on using natural materials, the German mattress brand Elsa has ranked highly in consumer evaluations since entering the Taiwan market in 2004. Established in Germany in 1924, Elsa has strived to satisfy customers for decades. The cold and damp winters in Taiwan make the island an ideal market for the diverse products Elsa manufactures, including woolen carpets, blankets, socks, stockings and nightclothes. The main competitor of Elsa in the Taiwan market is Simmons Mattress Company of the United States, which began mass producing spring beds in 1876 and built up a firm position in the Taiwan market. Simmons is the leader in its market niche, and focuses on eliminating the stress of shopping for mattresses by combining innovation with comfort. Simmons seeks to assure customers that they are purchasing a quality mattress, and also to convince them quality sleep is essential for dealing with the turbulence of daily life. For individuals who have trouble sleeping, Simmons offers quality mattresses that often exceed consumer expectations. Elsa faces a challenge in competing with a well entrenched rival like Simmons. Knowledge is a key concern. Rather than merely selling quality mattresses, Simmons educates consumers to maximize their sleeping experience and ensure a healthy lifestyle. In 2005, Elsa plans to team up with Teco Company and offer consumers a 50% discount off of an Elsa mattress when purchasing a Teco humidifier. Given the damp climate in Taiwan, consumers with a humidifier in their bedrooms should sleep more comfortably and healthily. Such a partnership should help improve Elsa's position in this highly competitive market.

Situation 2

Long term care emerged in Taiwan in the late 1980s. A turning point occurred in 1997 with the passage of the Senior Citizens Welfare Law, which placed unregistered healthcare institutes under pressure and eventually saw them abolished by 2000. This legislation ushered in the rapid growth of institutional-based organizations from 1998, with stable growth expected well beyond 2000. Competition among institutional-based long term care facilities currently is fierce. A recent market survey indicated that while only around 20% of all disabled elderly in Taiwan receive institutional-based care, 30% of the disabled elderly in Taiwan require such care (Department of Health, 1997). This discrepancy represents a market demand of at least 18,000 individuals. To meet this demand, Taiwan has relied on small-scale institutional care facilities. Such facilities have become popular for four reasons. First, family members without professional training account for 69% of all care providers for the disabled elderly in Taiwan. This situation creates high emotional and financial stress, and providing professional treatment to the disabled elderly can greatly alleviate family tensions. Second, the annual growth rate of disabled elderly in Taiwan is nearly 20%. Third, modern lifestyles and urbanization have significantly transformed familial patterns, as reflected by the tendency of adults to live apart from their parents and offer their parents less assistance than previously. Meanwhile, increasing daily pressures in daily life, family interactions, and the growing female workforce have reduced numbers of non-professional caretakers for the disabled elderly. Consequently, professional care givers are increasingly important for meeting demand. Fourth, a clear discrepancy in supply of institutional-based long term care facilities exists between urban and rural areas. While demand for such facilities is lower in rural areas, marketing opportunities still exist for smaller scale facilities.

Situation 3

Wound care treatment is crucial in nursing care, and involves the assessment of wound severity and appropriate treatment. Increasing life expectancies globally over the past three decades, a growing elderly population and the eradication or alleviation of many systemic diseases have all contributed to the urgent need to clinically treat patients with chronic illnesses, especially those with difficult to heal wounds. While striving to heal patients with wounds by adopting the latest therapeutic treatment strategies, medical personnel benefit not only patients, but also their relatives and society as a whole. Reducing wound healing times reduces fatality rates, inhibits disease growth during the early stages, alleviates the burden on relatives in terms of manpower and financial resources and reduces hospital expenditures, ultimately reducing already strained National Health Insurance resources. Statistics demonstrate the severity of this problem. In the United States alone, over 1 million patients seek treatment annually for chronic wounds, with treatment costs totaling several hundred million dollars. Expenses associated with length of hospital stay and the extent of wound care treatment are valuable indexes of the severity of the wound treatment problem. Thus, Taiwanese hospital administrators are increasingly emphasizing the need to reduce wound treatment associated costs in clinical practice. A recently developed wound management procedure, vacuum-assisted closure (VAC), applies negative pressure to a wound through a porous, open-cell foam that fills the wound cavity. The advantages include rapid wound healing, reduced pain, shorter hospital stays, lower medical costs and less need for nursing care. This procedure can also be applied to patients with multiple wounds, as well as to recurring wounds suffered by many elderly patients.

B

When was the German mattress brand Elsa established?

In 1924

When did Simmons Mattress Company of the United States begin mass producing spring beds?

In 1876

When does Elsa plan to team up with Teco Company and offer consumers a 50% discount off of an Elsa mattress when purchasing a Teco humidifier?

In 2005

C

What placed unregistered healthcare institutes under pressure and eventually saw them abolished by 2000?

The Senior Citizens Welfare Law

What currently is fierce?

Competition among institutional-based long term care facilities

What have reduced numbers of non-professional caretakers for the disabled elderly?

Increasing daily pressures in daily life, family interactions, and the growing female workforce

D

How have medical personnel benefited patients?

By striving to heal patients with wounds by adopting the latest therapeutic treatment strategies

How does vacuum-assisted closure (VAC) apply negative pressure to a wound?

Through a porous, open-cell foam that fills the wound cavity.

How are expenses associated with length of hospital stay and the extent of wound care treatment valuable?

They serve as indexes of the severity of the wound treatment problem.

E

Where do over 1 million patients seek treatment annually for chronic wounds, with treatment costs totaling several hundred million dollars?
In the United States

What are the benefits of reducing wound healing times?
It reduces fatality rates, inhibits disease growth during the early stages, alleviates the burden on relatives in terms of manpower and financial resources and reduces hospital expenditures, ultimately reducing already strained National Health Insurance resources.

How long have life expectancies been increasing globally?
Over the past three decades

F

Situation 1
1. A 2. C 3. A 4. C 5. B

Situation 2
1. B 2. C 3. B 4. A 5. A

Situation 3
1. C 2. C 3. B 4. C 5. B

G

Situation 1
1. C 2. C 3. A 4. B 5. B 6. A 7. C 8. A 9. B 10. B 11. A 12. A 13. C

Situation 2

1. B 2. A 3. A 4. C 5. C 6. A 7. B 8. B 9. A 10. C 11. A 12. A 13. C 14. B 15. B

Situation 3

1. C 2. B 3. C 4. A 5. C 6. C 7. B 8. C 9. B 10. A 11. B

I

Situation 4

Fast food items are immensely popular among Taiwanese, with instant noodles being no exception. Delicious, convenient, inexpensive and healthy, instant noodles have been a staple food item among Taiwanese for more than four decades. When production of instant noodles in Taiwan began in 1967, the International Food Company from Japan initially dominated the market. However, after Wei Lih Food Manufacturers established a food processing plant in Changhua in 1970, local producers began to gradually erode Japan's market dominance. Taiwanese manufacturers initially imitated Japanese products, but eventually they began making adjustments to appeal to local tastes, such as adding chicken essence to instant noodles and enclosing seasoning packets that included salt, monosodium glutamate, pepper and other flavorings. With other local enterprises entering the market, including Ve Wong Company, Uni-President Enterprises, Vedan Enterprise Corporation and even the King Car Group, local production of instant noodles gradually matured, and local products gradually captured the dominant market share. In 2002, after 37 years in business, Taiwanese manufacturers of instant noodles achieved revenues of approximately 3 billion New Taiwanese dollars. According to the 2003 Integrated Consumer Tendency (ICT) report on Taiwanese consumer trends, 15-29 year olds are the biggest consumers of instant noodles. Increasing market demand for diet food products has led to innovations in instant noodles.

Furthermore, Taiwan's recent entry into the World Trade Organization has created opportunities for technology cooperation aimed at better satisfying consumer tastes, enhancing production management practices and improving after-sales service. Given the above trends, local manufacturers of instant noodles face new opportunities and challenges.

Situation 5

The Taiwanese economy has grown strong during the past decade, and the average income has now reached $US 13,000. Improved living standards have made Taiwanese more health conscious and recreation-oriented. Although most employees undergo a routine physical examination annually, including blood tests, chest x-ray examinations and heart-lung function testing, such examinations do not accurately reflect the current condition of patients. Most individuals pay extra for an MRI examination when undergoing their routine physical examination. No longer restricted simply to identifying lesions, MRI examinations have become an effective means of determining the current status of human organs and vessels. One of the advantages of an MRI exam is that no prior preparations are necessary. Patients can eat normally, continue with their normal daily routines and continue taking any prescribed medications. Typically lasting from 20 to 45 minutes, depending on the information required by the physician, the procedure simply requires the patient to lie in a supine position and remain still. Patients can be accompanied by relatives in the scan room, and are closely supervised by medical technologists. Additionally, the magnetic chamber includes an intercom system should the patient require. A contrast agent may be administered to enhance the study, but no precautions are necessary. Patients are free to consult with the attending physician or medical technologists to discuss any concerns. Importantly, the examination involves no radiation, with data acquired via other means, which include axial, sagittal and coronal observations. Hospitals increasingly realize the potential of comprehensive physical examinations for generating revenue, thus reducing pressures on the already strained national health insurance system.

Situation 6

As a novel radiation therapy and planning system that can increase cure rates for cancer patients, tomotherapy offers the most advanced radiation delivery system available through its enhanced dose modulation and accurate targeting of specific locations. Tomotherapy allows physicians to verify treatment volumes in advance through 3D imagery via TomoImage scanning, ensuring that treatment fits a therapeutic strategy. Additionally, this system delivers helical tomotherapy to targets while minimizing damage to healthy tissue, thus optimizing dose delivery for all patients. Pioneered by Professor Thomas Rockwell Mackie and the mathematician and software engineer Paul J. Reckwerdt at the University of Wisconsin-Madison ten years ago, tomotherapy combines a treatment planning optimizer, a linear accelerator, computed tomography (CT) and a complex intensity modulation radiation therapy (IMRT). Among its unique features, tomotherapy offers precise planning through using a treatment planning optimizer that is easier to use than conventional treatment planning systems. Moreover, tomotherapy ensures precise positioning through using a unique verification CT to confirm the tumor position before each treatment, enabling precise delivery of the radiation dosage. Furthermore, tomotherapy also ensures precise delivery of the prescribed dosage to the intended area owing to its ability to combine complex IMRT with spiral delivery, thus concentrating the radiation on the tumor and depositing less radiation in surrounding healthy tissue. In sum, this therapeutic treatment system is widely anticipated to be adopted among hospital oncology departments to provide enhanced medical care for cancer patients.

J

Why did Taiwanese manufacturers begin making adjustments instead of just imitating Japanese products?
To appeal to local tastes

Why have innovations been made in instant noodles?
Because of increasing market demand for diet food products

Why have there been opportunities for technology cooperation aimed at better satisfying consumer tastes, enhancing production management practices and improving after-sales service?
Because of Taiwan's recent entry into the World Trade Organization

K

What is the average income in Taiwan?
$US 13,000

What have become an effective means of determining the current status of human organs and vessels?
MRI examinations

What do hospitals increasingly realize?
The potential of comprehensive physical examinations for generating revenue

L

How does tomotherapy optimize dose delivery for all patients?
By delivering helical tomotherapy to targets while minimizing damage to healthy tissue

How can physicians ensure that treatment fits a therapeutic strategy?
By verifying treatment volumes in advance through 3D imagery via TomoImage scanning

How does tomotherapy enable precise delivery of the radiation dosage?
By ensuring precise positioning through using a unique verification CT to confirm the tumor position before each treatment

M

What is a novel radiation therapy and planning system that can increase cure rates for cancer patients?

Tomotherapy

Who pioneered the use of tomotherapy?

Professor Thomas Rockwell Mackie and the mathematician and software engineer Paul J. Reckwerdt at the University of Wisconsin-Madison ten years ago

How does tomotherapy ensure precise positioning?

Through using a unique verification CT to confirm the tumor position before each treatment, enabling precise delivery of the radiation dosage

N

Situation 4

1. B 2. C 3. C 4. A 5. B

Situation 5

1. B 2. A 3. A 4. C 5. A

Situation 6

1. C 2. C 3. B 4. A 5. C

O

Situation 4

1. B 2. B 3. A 4. C 5. B 6. A 7. A 8. C 9. B 10. A 11. C

Situation 5

1. C 2. A 3. B 4. B 5. A 6. C 7. B 8. B 9. A 10. C 11. C 12. B 13. A 14. C

Situation 6

1. B 2. A 3. C 4. A 5. C 6. A 7. B 8. A 9. C 10. B

A

Situation 1

Despite representing a considerable portion of the game sector, on-line gaming in Taiwan lags behind Korea in terms of development, explaining the large trade imbalance between the two countries in this industry. Owing to the enormous potential of on-line gaming, the Taiwanese government has added this sector to the priority list of growth areas for national development. Current government policies aim not only to satisfy domestic demand for on-line games by replacing imported Korean products with innovative local products, but also to develop state-of-the-art globally competitive online games by emphasizing design, manufacturing and testing. Additionally, the government is encouraging two novel promotional strategies. First, fashionable trends and comic characters that appeal to the key market sector, 13-26 year olds, are adopted in game design and promotion. Second, new games are promoted together with current comic book series, movies and television programs to raise brand awareness. Although this strategy requires considerable time in investment to promote the connection between the new game and current comic book series, movies and television programs, it offers a means of design and marketing on-line games that appeal to Chinese consumers.

Situation 2

Before the Taiwanese government initiated the National Health Insurance Program in 1995, 13 different health insurance schemes covered nearly 60% of residents of Taiwan, with the remaining population left to pay for treatment entirely on their own. To effectively address this problem, the Council for Economic Planning and Development established a planning committee in 1988 to develop a single mandatory and universal health insurance program. The program took effect in March 1995 after the Legislative Yuan passed the National Health Insurance (NHI)

Act. The NHI Bureau was established to achieve three goals. First, the NHI Bureau sought to ensure that all Taiwanese residents received insurance coverage. According to the Department of Health of Taiwan, around 96% of all Taiwanese residents had NHI program coverage as of December 1997. Second, the NHI Bureau should improve medical care quality and increase competition among healthcare providers. Unlike under the previous system, under NHI, patients can freely select where they receive medical care. Third, by implementing a Global Budget payment system and introducing payment on a per case basis rather than a per-visit basis, the NHI Bureau could help control hospital health care costs and resource utilization efficiency. This approach enabled the economic provision of high quality health care services. In sum, the NHI program has expanded the size and scope of the medical care sector, increased market competitiveness, and created incentives for continuously improving operational efficiency.

Situation 3

When Real-Sun Information Technology Company entered the medical services sector in Taiwan eight years ago, less than 10% of all hospitals outsourced their information system needs. However, the establishment of the National Health Insurance scheme in 1995 impelled Taiwanese hospitals to computerize their operations. Moreover, governmental policy aimed at upgrading the computerized capabilities of hospitals attracted many information technology firms to this newly emerging market. Consequently, nearly 80 information technology companies invested in the medical information sector, and Internet companies followed them into the market, representing an initial investment of nearly $NT 20,000,000. Following several company mergers and considerable investment, Taiwan has around 20 domestic information technology firms involved in the medical sector. The customer base for this sector comprises 2600 clinics and 47 hospitals island wide. Real-Sun Information Technology Company has recently merged its personnel and resources with another medical software firm to increase its market share. As domestic medical health networks continue to evolve in this newly emerging market, 40 local firms have been established, while globally there are over 10,000 medical information-oriented websites. Several websites provide online research capabilities, providing access to electronic journals, disease-related information for diagnostic purposes, on-line queries, examinations, automatic registration functions, personal health tips and online ordering capabilities. Meanwhile, all Internet-based medical information companies can handle kinesiology and medicine classification-related queries. Concerns regarding how to protect client information in an IC card format make extracting information from IC card contents for commercial purposes extremely difficult. Real-Sun has therefore designed IC-based duplicate cards for medical treatment. As credit cards have become a standard payment method, these duplicate IC cards can be used for medical payments, enabling thousands of clinics to adopt a uniform method of payment processing.

B

Why do current government policies aim to replace imported Korean products with innovative local products?

In order to satisfy domestic demand for on-line games

Why are fashionable trends and comic characters adopted in game design and promotion?

To appeal to the key market sector, 13-26 year olds

Why are new games promoted together with current comic book series, movies and television programs?

To raise brand awareness

C

What did the NHI Bureau seek to ensure?

That all Taiwanese residents received insurance coverage

What percentage of all Taiwanese residents had NHI program coverage as of December 1997?

Around 96%

What has the NHI program expanded?

The size and scope of the medical care sector

D

How did governmental policy attract many information technology firms to this newly emerging market?

By aiming to upgrade the computerized capabilities of hospitals

How has Taiwan developed around 20 domestic information technology firms involved in the medical sector?

Through several company mergers and considerable investment

How has extracting information from IC card contents for commercial purposes become extremely difficult?

By addressing concerns regarding how to protect client information in an IC card format

E

When did Real-Sun Information Technology Company enter the medical services sector in Taiwan?

Eight years ago

What has governmental policy aimed at upgrading the computerized capabilities of hospitals attracted?

Many information technology firms to this newly emerging market

What does the customer base for this sector comprise?

2600 clinics and 47 hospitals island wide

F

Situation 1

1. C 2. A 3. B 4. B 5. A

Situation 2

1. C 2. C 3. B 4. A 5. A

Situation 3

1. C 2. A 3. B 4. C 5. B

G

Situation 1

1. B 2. C 3. B 4. A 5. C 6. A 7. A 8. B 9. C 10. C

Situation 2

1. C 2. B 3. C 4. A 5. C 6. B 7. B 8. A 9. C 10. B 11. A

Situation 3

1. B 2. C 3. A 4. A 5. C 6. B 7. C 8. B 9. A 10. C 11. B 12. A 13. C

|

Situation 4

As a non-invasive form of nuclear medicine, positron emission tomography (PET) is preferable to conventional image reconstruction, which requires filtered back projection software and maximum likelihood expectation maximization software that often produces poor quality images owing to the limited number of photons and the slow convergence of original image data. Pioneered by the mathematicians Hudson and Larkin in 1994, an image reconstruction method based on ordered subset expectation maximization software has been widely adopted in experimental investigations using PET, and involves a fast algorithm enhancing the slow convergence of conventional software. Given the above considerations, hospitals must replace conventional software to produce high quality clinical images. Owing to budgetary constraints, National Health Insurance (NHI) will not cover the expense of this service, making this area a promising source of revenue for hospitals with the necessary technological capacity. Patients that are unwilling to cover the expense of this PET examination themselves, can choose from other less advanced but still high quality procedures subsidized by NHI, such as ultrasound, multi-slice computer tomography or magnetic resonance imaging. A notable example of such unsubsidized services is the PET Center of Shin Kong Wu Ho-Su Memorial Hospital, which has served over 4,000 patients over the past two years, generating substantial revenues for the hospital, thus reducing its reliance on NHI for funding. Each patient pays roughly $US 1,500 for this PET examination. Besides enabling the nuclear medicine departments in Taiwanese hospitals to fill a profitable market niche, this PET software-based medical procedure can eventually replace conventional computer image reconstruction software. Enhanced image reconstruction computer software based on maximizing ordered subset expectations is highly promising in clinical efforts to provide high-quality PET images. Several hospitals are currently developing marketing strategies to make patients aware of such high quality medical treatments.

Situation 5

As an excellent therapy for lung cancer patients, tomotherapy not only enables precise image-guided intensity modulated radiotherapy, but also provides valuable information regarding tumor changes during radiotherapy. Tomotherapy thus represents the future of image-guided IMRT in cancer treatment. The Cancer Center of National Taiwan University (NTU) utilizes all available resources to combat cancer. Numerous investigations have demonstrated the potential of tomotherapy in many unexplored areas in radiation oncology. As a compact, cost effective and high precision radiation therapeutic treatment system, tomotherapy includes a primary beam shield that reduces the operational costs associated with room shielding, an on-board oncology CT system, a rapid inverse planning system with a built-in optimizer, a full network with DICOM input, and a built-in patient scheduler. Taking TomoImage scans before each treatment enables physicians to determine whether a tumor is shrinking. After four or five weeks of treatment, the treatment dosage can be decreased as the tumor size shrinks. Known as slice therapy because it derives from tomography or cross-sectional imaging, the tomotherapy system resembles a computed tomography system: the patient lies on a bench that moves continuously through a rotating ring gantry. The gantry is fitted with a linear accelerator, which delivers a fan beam of photon radiation as the ring turns. The couch movement combined with the gantry rotation mean that the radiation beam spirals around the patient. Tomotherapy delivers intensity modulated radiotherapy (IMRT) using a multileaf collimator. A recent advance in radiation treatment involves IMRT altering the size, shape and intensity of the radiation beam depending on tumor size, shape and location. Tomotherapy achieves better cancer patient survival rates than other therapeutic treatment systems. The Cancer Center at NTU plans to collaborate with two innovators in this field: Professor Thomas Rockwell Mackie, a leading medical physicist, and Paul J. Reckwerdt, an accomplished mathematician and software engineer. Moreover, Tomotherapy Incorporated holds 70 patented technologies in this area, providing a valuable source of information for further understanding advanced applications of the tomotherapy system.   Although some clinicians do not consider tomotherapy a mature technology, preliminary results for treating prostate and lung cancer are encouraging. The precision of tomotherapy offers therapeutic potential for cancer patients ineligible for radiotherapy.

Situation 6

Medical technology recently has rapidly evolved, as evidenced by the increasing number of technologies available to radiation oncology departments, for example radiation pharmaceuticals and linear accelerators. Given the increasing market demand for advanced cancer therapeutic treatment strategies, according to the National Science Council of the Republic of China, Taiwan, revenues from the radiation oncology sector in Taiwan ranged between $NT 500 million to 1 billion dollars in 2004. Given the large market potential and strong government backing, hospitals with radiation oncology departments have expressed strong interest in adopting the latest technological applications. When creating a new model in this intensely competitive market, instrumentation companies strive to efficiently use available resources and identify effective market strategies. As an effective market strategy for understanding this highly competitive market, the four-point based market strategy can help clinical radiation oncology departments equip management professionals with appropriate and efficient marketing policies. Comprising product, price function, accuracy and promotion, the four-point based marketing strategy focuses on technology differentiation, in which hospital administrators stress how their product lines differ from those of other hospital centers. The price strategy sets prices based on the prices of competing products. Since radiation oncology strongly emphasizes accuracy, related technologies require precise instrumentation. New products face an intensely competitive market and, thus, promotion strategies must stress the unique features that differentiate them from other products. Finally, it is important to use the most appropriate agent for marketing purposes. In sum, the four-point based marketing strategy can help prepare radiation oncology departments in Taiwan for an intensely competitive global market that emphasizes state-of-the-art medical instrumentation and professionalism.

J

What does conventional image reconstruction require?

Filtered back projection software and maximum likelihood expectation maximization software that often produces poor quality images owing to the limited number of photons and the slow convergence of original image data

What has been widely adopted in experimental investigations using PET, and involves a fast algorithm enhancing the slow convergence of conventional software?

An image reconstruction method based on ordered subset expectation maximization software

What is positron emission tomography (PET) preferable to?

Conventional image reconstruction

What must hospitals must do to produce high quality clinical images?

Replace conventional software

What is highly promising in clinical efforts to provide high-quality PET images?

Enhanced image reconstruction computer software based on maximizing ordered subset expectations

K

Why is Tomotherapy Incorporated a valuable source of information for further understanding advanced applications of the tomotherapy system?
Because it holds 70 patented technologies in this area

Why is tomotherapy known as slice therapy?
Because it derives from tomography or cross-sectional imaging

Why is the gantry fitted with a linear accelerator?
To deliver a fan beam of photon radiation as the ring turns

L

How have hospitals with radiation oncology departments been able to express strong interest in adopting the latest technological applications?
Because of the large market potential and strong government backing

How can instrumentation companies create a new model in this intensely competitive market?
By striving to efficiently use available resources and identify effective market strategies

How can the four-point based market strategy help clinical radiation oncology departments equip management professionals with appropriate and efficient marketing policies?
By allowing them to understand this highly competitive market

M

What evidence supports that the medical technology has rapidly evolved recently?

The increasing number of technologies available to radiation oncology departments

What is the range of revenues generated from the radiation oncology sector in Taiwan?

Between $NT 500 million to 1 billion dollars

Why have hospitals with radiation oncology departments expressed strong interest in adopting the latest technological applications?

Because of the large market potential and strong government backing

N

Situation 4

1. C 2. A 3. C 4. A 5. C

Situation 5

1. B 2. B 3. A 4. C 5. B

Situation 6

1. B 2. A 3. C 4. B 5. C

O

Situation 4

1. C 2. C 3. B 4. A 5. A 6. C 7. B 8. B 9. A 10. C

Situation 5

1. B 2. A 3. A 4. C 5. C 6. A 7. B 8. B 9. A 10. C 11. B 12. A
13. C 14. B 15. A 16. C

Situation 6

1.B 2.A 3.C 4.A 5.B 6.A 7.C 8.A 9.A 10.C 11.B

A

Situation 1

The Taiwan operations of the German mattress brand "Elsa" were set up in 2004, with aspirations of servicing the entire Asia Pacific region. Following careful consideration and extensive marketing research, Elsa selected Taiwan as its retail and distribution hub in the Asia Pacific region. The Taiwanese bedding market recently has experienced sales growth, with statistics from the Directorate General of Budget, Accounting and Statistics, Executive Yuan revealing revenue growth from NT$ 24,000,000,000 in 1991 to NT$ 59,000,000,000 in 2001. Besides manufacturing wool and cotton mattresses in Germany, in Malaysia, Elsa also produces emulsion products such as pillows and mattresses. The excellent emulsion quality and relatively inexpensive labor in Malaysia have allowed Elsa to reduce overhead costs and pass these savings on to consumers in the form of competitive retail prices. Elsa has been operating in Asia for less than a year, and has sought to emphasize product and service quality via a segmentation marketing strategy involving another brand. Notably, Elsa views the production of recyclable mattresses as its social responsibility. Elsa has recently established three divisions: accounting, business and programming. The accounting division coordinates finances, factory operations and logistics, the business division focuses on sales and after sales service, and the programming division devises and implements marketing strategies, as well as coordinating art design projects. As an administrator in the programming division, Sheila is responsible for new employee orientation and on-the-job training, customer relations management-related practices and marketing strategy design.

Situation 2

The result of a joint investment by Taiwan Secom Company and Goldsun Development & Construction Company totaling NT$ 495,000,000, Goldsun Express

& Logistics (GE&L) Company (formerly Kuo Hsing Transport) was established in 1993 to undertake advanced and fully automated logistics. With a licensed customs office since 2002, the company has ample skilled staff from multi-disciplinary fields, including security, information technology, and construction. To provide comprehensive logistics services and achieve complete customer satisfaction, GE&L is striving to become an integrated and intelligent distribution center with exceptional customer service. Well positioned to lead the Taiwanese logistics service sector, GE&L bases its success on state-of-the-art distribution procedures, information technology, automated facilities, highly skilled industrial engineering professionals, third party logistics services, full-range B2B logistics solutions and sound supply-chain management practices. The company deals in hi-tech products, 3C products, entertainment media (DVD, VCD, VHS), cosmetics and pharmaceuticals, all of which are products for which enterprises tend to outsource distribution services. GE&L has a strong organizational framework, within which logistics professionals and IT consultants jointly provide a comprehensive and diverse range of logistics services. In these joint efforts, the three-stage client service approach is adopted to ensure customer satisfaction by tailoring offered services based on client needs. This approach uses EIQ analysis (commonly known as volume and variation analysis) to closely monitor the flow of incoming and outgoing goods. Additionally, analyses are performed, including analysis of the industrial characteristics of goods, business transaction requirements, environmental resource planning (ERP), parameter settings and process confirmation for large ERP systems, and linear motion testing. This procedure aims to minimize the influence of logistics services on business transactions during the initial service stage. Furthermore, the three-stage client service approach also considers how controlling inbound or outbound goods and inventory fully complies with the internal control system, ensuring that the logistic needs of customers are managed efficiently.

Situation 3

As Taiwan's highest health authority, the Department of Health (DOH) of the Executive Yuan is responsible for administering, supervising and coordinating local health agencies. DOH contributes to maintaining the health of Taiwanese residents by promulgating health-related measures to expedite the provision of convenient and efficient medical services. Promoting a healthy living space in which individuals can fully access quality healthcare services and accurate medical information enables all individuals to take increased responsibility for their well being. Key strategies implemented to achieve the above objective include establishing a National Health Insurance Review Committee to provide guidelines for improving the National Health Insurance system, promoting organizational reengineering by establishing the Bureau of Health Promotion and Bureau of Hospital Management, advancing national health education by establishing the National Health Education Research Group, setting up a modern health care network, enhancing health care for women and the disadvantaged, strengthening emergency medical care, encouraging diversification of hospital operations and, finally, implementing an information-based, nationwide system for preventing communicable diseases. To implement the above strategies, the Hsinchu General Hospital operates 18 outpatient services with a staff of 800, based on the concepts of quality, efficiency and family values. To achieve the objectives of promulgating governmental health care policy and offering quality medical health services to promote physical and mental well being, Hsinchu General Hospital has a bed capacity of 684, with 84 intensive care beds and 600 beds for patients with acute chronic illnesses.

B

Why was the Taiwan operations of the German mattress brand 'Elsa' set up in 2004?

To service the entire Asia Pacific region

Why has Elsa sought to emphasize product and service quality via a segmentation marketing strategy involving another brand?
Because it has been operating in Asia for less than a year

Why is Sheila responsible for new employee orientation and on-the-job training, customer relations management-related practices and marketing strategy design?
Because she is an administrator in the programming division

C

How is GE&L well positioned to lead the Taiwanese logistics service sector?
Owing to its success on state-of-the-art distribution procedures, information technology, automated facilities, highly skilled industrial engineering professionals, third party logistics services, full-range B2B logistics solutions and sound supply-chain management practices

How does GE&L have a strong organizational framework?
Its logistics professionals and IT consultants jointly provide a comprehensive and diverse range of logistics services.

How can GE&L ensure that the logistic needs of customers are managed efficiently?
Because its three-stage client service approach also considers how controlling inbound or outbound goods and inventory fully complies with the internal control system

D

What is the result of promoting a healthy living space in which individuals can fully access quality healthcare services and accurate medical information?
All individuals can take increased responsibility for their well being.

What do key strategies implemented to achieve the above objective include? Establishing a National Health Insurance Review Committee to provide guidelines for improving the National Health Insurance system, promoting organizational reengineering by establishing the Bureau of Health Promotion and Bureau of Hospital Management, advancing national health education by establishing the National Health Education Research Group, setting up a modern health care network, enhancing health care for women and the disadvantaged, strengthening emergency medical care, encouraging diversification of hospital operations and, finally, implementing an information-based, nationwide system for preventing communicable diseases

What does Hsinchu General Hospital do to achieve the objectives of promulgating governmental health care policy and offering quality medical health services to promote physical and mental well being?
It has a bed capacity of 684, with 84 intensive care beds and 600 beds for patients with acute chronic illnesses.

E

How does DOH contribute to maintaining the health of Taiwanese residents?
By promulgating health-related measures to expedite the provision of convenient and efficient medical services

How has DOH promoted organizational reengineering?
By establishing the Bureau of Health Promotion and Bureau of Hospital Management

How has DOH advanced national health education?
By establishing the National Health Education Research Group

F

Situation 1

1. B 2. C 3. B 4. A 5. B

Situation 2

1. C 2. B 3. A 4. C 5. A

Situation 3

1. C 2. B 3. C 4. A 5. C

G

Situation 1

1. B 2. B 3. C 4. A 5. C 6. B 7. C 8. A 9. B 10. A

Situation 2

1. C 2. B 3. C 4. A 5. A 6. B 7. A 8. C 9. A 10. B 11. B

Situation 3

1. B 2. B 3. A 4. C 5. A 6. C 7. A 8. C

I

Situation 4

With more than a century of history, National Taiwan University Hospital (NTUH) symbolizes the evolution of the Taiwanese medical care sector. The hospital has two main buildings, and the western wing was the largest and most modern hospital in Southeast Asia upon its construction in 1898. Committed to nurturing medical talent and developing outstanding research capabilities, NTUH strives to set precedents for the entire medical community. Given the inestimable value of life and the growth in global health consciousness, NTUH is globally renowned as a premier teaching hospital in Asia. Regarding the organizational structure of NTUH, decision making is performed by the secretarial department. Meanwhile, the administrative department comprises the financial, personnel, accounting, information technology and other smaller units. Furthermore, the medical departments include the nutritional, pharmaceutical, nursing and aesthetic units, as well as the hepatitis research center. The medical departments also include various units within the hospital, including three major departments involved in medical radiology. First, the nuclear medicine department provides radiological images for diagnosis by injecting radiological pharmaceuticals and using advanced methods such as PET. Second, the medical imaging department produces general diagnostic images using x-ray MRI examinations. Third, the cancer therapy center provides therapies such as surgery, chemotherapy and radiation therapy. NTUH is recognized as an international leader in several fields, especially hepatitis, organ transplantation, nasal and paralegals cancer, liver and stomach cancer therapy and antivenin research. As a teaching hospital, NTUH stresses education and research as well as medical services. In terms of education, the hospital seeks to foster medical talent in various fields by following a strong medical curriculum. As for research, NTUH integrates the resources of various fields and adopts state-of-the-art equipment to yield optimal research results. Regarding medical services, the hospital focuses on serving patient needs and strengthening its organizational structure. Besides offering nursing home care services, NTUH actively encourages international cooperation to remain abreast of global trends in medicine.

Situation 5

Committed to advancing the clinical diagnostics sector in Taiwan, Oxford Biosensors Corporation adopts a multidisciplinary approach, employing leading research scientists in electronics, materials science, electrochemistry and enzyme technology. Internationally recognized experts comprise the corporate advisory board, and the corporation draws heavily upon the personnel and resources of Oxford University, which is renowned worldwide for its innovations in electrochemistry and biosensor development. Through successfully commercializing technologies developed in academia using funding from global investment sources, Oxford Biosensors has generated a strong and growing intellectual property portfolio, based on a commitment to product development and broad expertise in biosensor design, development and manufacturing. Specifically, Oxford Biosensors provides medical diagnostic laboratories with accurate, efficient and inexpensive devices, and is a leader in this field. Effective healthcare requires accurate data regarding certain biochemical parameters, meaning that biosensors play a crucial role in modern healthcare owing to their specificity, miniaturized size, rapid response and relative inexpensiveness. Bioanalytical procedures are widely anticipated to be increasingly adopted for measuring metabolites, blood cations and gases. Established in 2000 based on technological innovations by Oxford University, the pilot plant manufacturing facility of Biosensors is located in nearby Yarnton, England. As a rapidly growing healthcare diagnostics development company, Oxford Biosensors provides a challenging and stimulating work environment for entrepreneurial individuals committed to constant technological innovation. Company employees have a strong sense of how they can contribute by closely collaborating on results-oriented endeavors. Oxford Biosensors focuses mainly on electrochemistry and enzyme technology. The electrochemistry program is particularly noteworthy as significant resources have been invested in commercializing biosensor technologies. The technical breakthroughs pioneered at Oxford Biosensors include the development of a diverse range of low complexity, hand-held diagnostic products. For example, the Multisense Dry Enzyme System satisfies consumer demand for rapid and accurate multi-parameter analysis by instantly providing essential diagnostic results.

Situation 6

Established in 1977, Sinphar Pharmaceutical Corporation became the first pharmaceutical manufacturer in Taiwan to receive both the National Award for Small and Medium-sized Enterprises and ISO 9001 accreditation in 1996. Later, in 2002, Sinphar was the only pharmaceutical manufacturer to receive the National Biotechnology & Medical Care Award for both its factory operations and product innovation. After investing over NT$ 230,000,000, Sinphar constructed the first research and development center in Taiwan to receive ISO 17025 and GLP accreditation. Licensed for operations in July of 2002, with production beginning in October of that year, the 12,000 square feet, five-story facility became a model for integrating domestic and international biotechnology research. An R&D budget of $NT 80,000,000 for 2003, along with further investment of up to $NT 450,000,000 over the next five years demonstrates the commitment of the Sinphar Group to developing state-of-the-art product technologies. Committed to continuously upgrading business operations, expanding upon patented technological innovations, and ensuring superior product quality, Sinphar utilizes its biotechnology expertise to extract and purify traditional Chinese medicine components in order to create new products for clinical trials and eventual commercialization. A state-of-the-art R&D Center greatly facilitates product technology research. A notable example is its clinical trial facility for Chinese herbal medicine, which applies advanced scientific approaches to extract and purify components of Chinese herbal medicine and, eventually, to standardize formulations for clinical studies.

J

What is NTUH globally renowned as?

A premier teaching hospital in Asia

What does the administrative department comprise?

The financial, personnel, accounting, information technology and other smaller units

In what fields is NTUH recognized as an international leader in?

Hepatitis, organ transplantation, nasal and paralegals cancer, liver and stomach cancer therapy and antivenin research

K

How is Oxford Biosensors a leader in this field?

By providing medical diagnostic laboratories with accurate, efficient and inexpensive devices

How do biosensors play a crucial role in modern healthcare?

Owing to their specificity, miniaturized size, rapid response and relative inexpensiveness

How do company employees have a strong sense of how they can contribute to Oxford Biosensors?

By closely collaborating on results-oriented endeavors

L

When did Sinphar receive the National Biotechnology & Medical Care Award for both its factory operations and product innovation?

In 2002

When did Sinphar's research and development center in Taiwan become licensed for operations?

In July of 2002

When did Sinphar's research and development center in Taiwan begin production?
In October of that year

M

What did Sinphar Pharmaceutical Corporation become the first pharmaceutical manufacturer in Taiwan to receive?
Both the National Award for Small and Medium-sized Enterprises and ISO 9001 accreditation

Why did Sinphar Pharmaceutical Corporation receive the National Biotechnology & Medical Care Award?
For both its factory operations and product innovation

How much did Sinphar invest to construct the first research and development center in Taiwan?
Over NT$ 230,000,000

N

Situation 4
1. B 2. A 3. C 4. B 5. C

Situation 5
1. B 2. B 3. C 4. A 5. A

Situation 6
1. C 2. B 3. B 4. A 5. C

O

Situation 4

1. B 2. A. 3. C 4. C 5. A 6. B 7. B 8. C 9. A 10. C 11. A 12. A

13. B 14. C 15. B 16. A

Situation 5

1. B 2. B 3. A 4. C 5. A 6. A 7. C 8. C 9. B 10. B 11. A 12. C 13. A

Situation 6

1. B 2. B 3. C 4. A 5. B 6. C 7. A 8. B 9. B 10. A

Answer Key
Introducing a Division or a Department
組或部門介紹

A

Situation 1

The Educational Training and Orientation Department at China Motors Corporation (CMC) integrates the efforts of human resources personnel, training instructors and library employees. The Department performs specific activities to continuously upgrade the professional skills of company employees through a comprehensive education and training curricula. First, while closely following the missions and planning strategy of CMC, the Department assesses the talents of employees and recommends ways of better utilizing their professional skills and potential to more effectively respond to the intensely competitive and constantly changing market. Second, the Department maintains an employee training center to fully orient new employees regarding company operations and equip them with the professional skills necessary to perform daily tasks. The center often consults with its counterparts in industry and government regarding how to improve its operations. Third, the Department maintains a consultation room that allows colleagues to freely discuss professional and personal concerns regarding how to better fulfill their career aspirations while maintaining a healthy family and personal life. Finally, the Department maintains a well stocked library containing numerous books, periodicals and multimedia teaching materials, thus encouraging employees to continuously upgrade their reading skills to maintain their professional skills. Moreover, the Department implements an apprenticeship model in which new employees can work in specific areas under the guidance of more seasoned colleagues, thus fostering an organizational environment of mutual respect.

Situation 2

In line with corporate goals of enhancing biotechnology product quality, reducing overhead costs and adopting the latest professional technologies, the Sales Department at Taiwan Sugar Corporation (TSC) has adopted an integrated marketing approach that stresses product differentiation. Advanced biotechnology products are widely anticipated to reach commercialization within the next two years. The Biotechnology Division of TSC focuses on developing product technologies and technical expertise in fermentation, extraction, biological reagents, biomedical materials, biopharmaceuticals and in vitro diagnostics. The Division is preparing to install extraction equipment for water and organic solvents, operating lines and packing facilities to manufacture various high value-added products that adhere to FDA standards. Since research and development, procurement, production, sales and technical services are all integral to corporate success, the organizational structure of TSC is both flexible and highly responsive to market fluctuations. Given the recent emergence of biotechnology, TSC has invested heavily in developing biotechnology-related technologies to meet expected strong future market demand, as demonstrated by the Taiwanese government policy of encouraging the local private sector to enter this field. The Sales Department adopts flexible procedures for determining the core competence and commercial viability of TSC in the biotechnology market, especially in functional foods and cosmetics. Moreover, the Department is committed to educating customers regarding the health benefits of TSC's biotechnology products.

Situation 3

In existence for a decade, the Service Division of Goldsun Express & Logistics Corporation (formerly Kuo Hsing Transport) comprises ten employees dedicated to devising and implementing marketing strategies, as well as coordinating customer design projects. The smallest division in the company in terms of employees, the Service Division also operates a service center, staffed by five service engineers and an engineering assistant, all with undergraduate degrees related to their areas of experience and an average of two years professional experience in the logistics sector.

The Service Division strives to give Goldsun Express & Logistics Corporation a competitive edge in Taiwan's intensely competitive logistics sector by focusing on customer service, quality management, service quality and customer satisfaction. Although not directly generating revenue for the company, the Division is still indispensable in daily operations, integrating the efforts of other divisions to execute a cohesive business strategy based on customer relations management-related practices. For instance, logistics services require punctual product delivery to satisfy customers. In terms of daily operations, the Service Division handles electronic-based orders, current accounts, delivery, stock inventory control, stock inventory-related data and delivery of hi tech products to wafer fabs and retail shops. Products include 3C products such as mobile phones, personal computers (PC), PC peripherals, personal digital assistants (PDA), PDA peripherals and entertainment media products, including DVDs, VCDs, CDs and VHS tapes. The Division also supervises delivery of cosmetics and pharmaceutical products to regional hospitals, medical centers and department stores.

B

How does the center attempt to improve its operations?
By consulting with its counterparts in industry and government regarding how to improve its operations

How does the consultation room that the Department maintains help colleagues?
It allows them to freely discuss professional and personal concerns regarding how to better fulfill their career aspirations while maintaining a healthy family and personal life.

How does the Department encourage employees to continuously upgrade their reading skills to maintain their professional skills?

By maintaining a well stocked library containing numerous books, periodicals and multimedia teaching materials

C

Why has the Sales Department at Taiwan Sugar Corporation (TSC) adopted an integrated marketing approach that stresses product differentiation?

To stay in line with corporate goals of enhancing biotechnology product quality, reducing overhead costs and adopting the latest professional technologies

Why is the Biotechnology Division preparing to install extraction equipment for water and organic solvents, operating lines and packing facilities?

To manufacture various high value-added products that adhere to FDA standards

Why is the Division preparing to install extraction equipment for water and organic solvents, operating lines and packing facilities?

To manufacture various high value-added products that adhere to FDA standards

Why is the organizational structure of TSC both flexible and highly responsive to market fluctuations?

Because research and development, procurement, production, sales and technical services are all integral to corporate success

Why has TSC invested heavily in developing biotechnology-related technologies?

To meet expected strong future market demand

D

What is the smallest division in Goldsun Express & Logistics Corporation in terms of employees?

The Service Division

What does the Service Division strive to do?
Give Goldsun Express & Logistics Corporation a competitive edge in Taiwan's intensely competitive logistics sector

What do logistics services require?
Punctual product delivery to satisfy customers

What does the Service Division handle in daily operations?
Electronic-based orders, current accounts, delivery, stock inventory control, stock inventory-related data and delivery of hi tech products to wafer fabs and retail shops

What does the Division supervise?
Delivery of cosmetics and pharmaceutical products to regional hospitals, medical centers and department stores

E

How long has the Service Division of Goldsun Express & Logistics Corporation been in existence?
For a decade

How many service engineers does the Service Division staff?
Five

How much professional experience do the service engineers have in the logistics sector?
An average of two years

F

Situation 1

1. C 2. A 3. C 4. B 5. A

Situation 2

1. C 2. A 3. B 4. B 5. A

Situation 3

1. B 2. C 3. A 4. C 5. C

G

Situation 1

1. C 2. A 3. B 4. C 5. A 6. A 7. B 8. B 9. C 10. B 11. A

Situation 2

1. B 2. B 3. A 4. C 5. C 6. A 7. C 8. B 9. C 10. A

Situation 3

1. C 2. A 3. C 4. C 5. A 6. B 7. B 8. A 9. C 10. A

I

Situation 4

Established in 1988 as a producer of quality motors, Yuan Electrical Machinery Company has accumulated extensive experience in developing high-accuracy instrumentation and advanced machinery to meet diverse consumer needs. The company closely collaborates with microcomputer machinery manufacturers and machinery calibrators to achieve accuracy and precision. Additionally, the company adopts a closely supervised quality control inspection system in manufacturing to ensure that its motors have high torsion, low noise emissions and long life. Machinery is exported mainly to Germany, Hong Kong, Japan, the United States and other southeast Asian countries. Led by a chief administrator in charge of quality control, an administrative staff of 30 employees focuses on five to six quality control-related items. Group leaders coordinate the efforts of employees in their groups by encouraging strong cooperation through constant training to drill members on independent problem solving skills. The Administrative Department is mainly staffed by university graduates with backgrounds in quality control and related fields, with several also having masters and doctoral degrees in fields related to enhancing product quality. The Department recruits primarily from nearby universities and technological institutes, offering recent university graduates numerous management opportunities. While adopting a global perspective towards manufacturing quality products, the department focuses on the following directions: rigorously controlling manufactured product quality to satisfy consumer demand, adopting advanced technologies to gain consumer confidence, emphasizing customer needs to ensure consumer expectations are satisfied and approaching management with a sense of community service. Particularly, the Department emphasizes product quality by adopting the latest technological practices and stressing workplace safety. Demonstrating its commitment to excellence, the department contributed crucially to the company receiving ISO 9000 and 9002 product certification. Furthermore, the Department is striving to achieve compliance with ISO 14000 standards, something which requires a thorough understanding of design quality and quality control to maintain sound manufacturing practices.

Situation 5

Cancer has been the leading cause of death in Taiwan since 1982. Committed to providing advanced research capabilities and expertise in cancer therapy, the Cancer Center of National Taiwan University Hospital was established in 1999, and comprises a chemotherapy department, radiation therapy department, nuclear medicine laboratory, clinical experiment laboratory, radiology biological laboratory, biological statistical laboratory, outpatient services for tumor victims, a chemotherapy treatment room, and a ward for tumor patients. The Radiation Therapy Department is the heart of the Cancer Center and focuses on four treatment areas: a state-of-the-art IMRT-linear accelerator, Co-60 therapeutic machinery, a computer tomography simulator and a block-making room. While providing a diverse range of services for cancer patients, the Cancer Center provides a customized treatment strategy that incorporates the latest treatments in surgical oncology, radiation oncology and chemotherapy. The center has highly talented staff, including radiation oncologists, surgical oncologists, medical oncologists, diagnostic radiologists, operate advanced instrumentation such as a linear accelerator, CO-60 and CT, as well as medical physicists who plan therapy. These talented staff help the Cancer Center to achieve the following objectives: providing patient-focused healthcare to enhance patient quality of life, satisfying individual patient requirements and those of their relatives in a nurturing environment, fostering innovative research to prevent, detect and treat various cancers in their early stages, and training future cancer treatment experts and researchers. Additionally, while conducting preliminary and clinical research, the Cancer Center nurtures professional talent in therapeutic, physiological and psychological aspects of cancer research. Moreover, nurses receive specialized training to care for radiation therapy patients, including help such patients cope with the discomfort caused by adverse skin effects and gland deficiencies. The Cancer Center focuses particularly on breast, nasal and pharyngeal, and lung cancers. The center also provides brachytherapy for cervical, ovarian and bladder cancers. Notably, the enhanced form of brachytherapy offered at the center achieves 5-year survival in 95% of patients with superficial tumors. Besides providing state-of-the-art medical services for patients, the Center also offers support for patients and relatives coping with cancer therapy.

Situation 6

Established in 1967, the Radiology Technology Department at Cheng Hsin Medical Center has provided nearly five decades of service, and has trained many outstanding radiologists who are now spread throughout Taiwan. The Medical Imagery Department is divided into radiology, nuclear medicine and radiotherapy sections, specializing in x-ray, CT and MRI examinations, respectively. The radiology unit has state-of-the-art facilities and highly skilled personnel to provide prompt and efficient care. The radiology team continuously pursues advanced technology applications to offer continually improving therapeutic services. With its excellent specialists and facilities, the radiology unit has not only received accreditation from the Chinese Society of Radiology, but also operates a residential training program under the auspices of the Department of Health. The specialized equipment of the unit includes a high-tech magnetic resonance imaging system, rapid computerized tomography, a dual energy x-ray bone densitometer, mammography, high-resolution ultrasonography and digital subtraction angiography. Nuclear medicine provides imaging and functional evaluations for various body organs, as well as radioimmunoassay analysis, including nuclear cardiology and neuropsychiatry. A research project is currently underway on administering radioactive iodine 13 1 to treat patients with thyroid disorders as a part of long-term follow-up therapy. The radiotherapy unit is fitted with a double density linear accelerator for treating patients suffering from cancer, supported by a high precision radiographic simulator, for accurately detecting tumor size and location. Driven by a high-powered computer, the therapeutic planning system can determine accurate dosages for therapeutic purposes. This service prioritizes quality control in therapeutic planning. As for future trends, the hospital is digitalizing all of its medical images, and the PACS system provided the entire hospital with an image transmissions system in June of this year.

J

What does the company do to ensure that its motors have high torsion, low noise emissions and long life?

It adopts a closely supervised quality control inspection system in manufacturing.

What do group leaders do to encourage strong cooperation through constant training to drill members on independent problem solving skills?

They coordinate the efforts of employees in their groups.

What has the department contributed crucially to?

The company receiving ISO 9000 and 9002 product certification

K

How long has cancer been the leading cause of death in Taiwan?

Since 1982

How long has the Cancer Center of National Taiwan University Hospital been operating?

Since 1999

How is the Radiation Therapy Department the heart of the Cancer Center?

It focuses on four treatment areas: a state-of-the-art IMRT-linear accelerator, Co-60 therapeutic machinery, a computer tomography simulator and a block-making room.

How does the Cancer Center incorporate the latest treatments in surgical oncology, radiation oncology and chemotherapy?

By providing a customized treatment strategy

How does the Cancer Center help patients and relatives cope with cancer therapy?
By offering support

L

Why does the radiology team continuously pursue advanced technology applications? To offer continually improving therapeutic services

Why has the radiology unit has received accreditation from the Chinese Society of Radiology?
Because of its excellent specialists and facilities

Why can the therapeutic planning system determine accurate dosages for therapeutic purposes?
Because it is driven by a high-powered computer

M

When was the Radiology Technology Department at Cheng Hsin Medical Center established?
in 1967

How is the Medical Imagery Department divided?
Into radiology, nuclear medicine and radiotherapy sections, specializing in x-ray, CT and MRI examinations, respectively

How does the radiology unit provide prompt and efficient care?
Through its state-of-the-art facilities and highly skilled personnel

N

Situation 4

1. C 2. A 3. C 4. B 5. C

Situation 5

1. A 2. C 3. C 4. B 5. A

Situation 6

1. C 2. B 3. B 4. A 5. C

O

Situation 4

1. B 2. C 3. A 4. C 5. A 6. C 7. C 8. B 9. C 10. A 11. A 12. C

Situation 5

1. B 2. C 3. C 4. A 5. C 6. B 7. A 8. C 9. C 10. B 11. B

Situation 6

1. C 2. C 3. A 4. C 5. B 6. B 7. C 8. A 9. A 10. B 11. B 12. B

Answer Key
Introducing a Technology
科技介紹

A

Situation 1

China Motors Corporation (CMC) recently constructed a project documentation and version control (PDVC) system, thus increasing its productivity and enhancing its competitive edge in the domestic automotive sector. Previously, new automobiles were spray painted on the factory floor, an inefficient and time consuming process. The advanced PDVC system improves automation control technologies by foreseeing potential production bottlenecks. Whereas global automakers generally produce cars in large quantities, CMC produces a smaller quantity of more diverse products. Demonstrating its success, the PDVC system has reduced personnel numbers on the factory floor from fifteen to six, while reducing wasted storage capacity from 68,071 to 55,031 units, saving CMC NT$ 2,000,000 over two years. Catering to individualized consumer tastes requires the automobile industry to effectively manage manufacturing procedures to simultaneously optimize factory floor use and satisfy consumer demand. The PDVC system helps to achieve this by arranging the simultaneous production of various models, striving to fill factory orders, and making on-line queries regarding how to increase upper and lower supply chain efficiency. Such technological innovations will ultimately improve the global competitiveness of the Taiwanese automotive sector. Technological developments in this sector focus mainly on flexibility in production scheme, online production and further integration of information technologies and, ultimately, result in a fully automated and optimally managed factory floor.

Situation 2

Elevated living standards and strong consumer demand in Taiwan have fueled technological advances in manufacturing liquid crystal displays. Despite having successfully developed technology for producing medium and full sized liquid crystal color displays, AU Optronics Corporation has yet to develop the equivalent technology for extremely large sized displays, primarily owing to the prohibitive costs. Until the 1990s, liquid crystal displays could not simultaneously achieve a competitive price, large size and high quality. The main applications of liquid crystal displays are in notebooks and cellular phones. AU Optronics Corporation mainly produces large-sized panels for use in desktop monitors, notebook PCs, LCD TVs and other consumer appliances. The company's retail sales of such panels reached NT$14,596 million for March 2005, with shipments reaching 2.33 million pieces during the same month, increased 29.1% from the previous month. Further demonstrating the phenomenal growth of AU Optronics recently, shipments of small- and medium-sized panels increased by 23.4% to reach 3.39 million units during March 2005. With the merger of Acer Display Technology Corporation (a subsidiary of Acer) and Unipac Optoelectronics Corporation (a subsidiary of United Microelectronics Corporation) in 2001, AU Optronics obtained the most advanced TFT-LCD technologies and became known as a one-stop-shop capable of providing both large and small-medium applications. By the end of 2003, the operations of AU were spread across China, Europe, Japan, South Korea, Taiwan and the United States, the company had over 20,000 employees, and annual revenues had exceeded $US 3 billion. Notably, the ability of the corporation to integrate life sciences and medical technology has resulted in state-of-the-art photoelectronic laser technology applications. Given its technological capabilities, AU Optronics is forging ahead into OLED, LPTS, MVA and transflective technologies. Product innovation has created added value and increased efficiency, leading to sustained growth in corporate profits and enhancing TFT-LCD performance, as demonstrated by the 656 local and international patents the company owns. The effective integration of multidisciplinary fields such as optoelectronics, mechanical engineering, electrical engineering and materials science will enable AU Optronics to continue to succeed.

Situation 3

Liver cancer has ranked as a leading cause of mortality in Taiwan during the past decade because of Taiwan's unusually high incidence of Hepatitis B and Hepatitis C. Taiwan has over 10,000 fatalities annually from chronic liver hepatitis, cirrhosis or liver cancer, and 80 to 90% of those affected carry the Hepatitis B or Hepatitis C viruses. Individuals with Hepatitis B have a 150 times greater likelihood of contracting liver cancer than those without the virus, explaining the high levels of hepatitis associated liver cancer among Taiwanese. Five major therapeutic methods are available for treating liver cancer: surgery, including either therapeutic surgery or organ transplantation; blockage of blood flow, such as embolism of the liver artery; chemotherapy, including treatment for the entire body and interarterial injection; local injection, including injection of alcohol and glacial acetic acid; and temperature therapy, including microwave therapy, high temperature therapy and radiation therapy. Radiation therapy is the most extensively adopted of these therapeutic methods. Advances in digital technology have led to 3D computer-animated control stations being widely adopted to treat various cancers using radiation therapy. Radiation therapy has advanced rapidly during the past decade: from CO-60 to 3D conformal therapy. Previously, surgery was the only therapeutic method of treating liver cancer. While the one-year survival rate following surgery can reach 80%, the five year survival rate is only 50%. Ultimately, survival rates depend on factors such as tumor position, size, degree of cirrhosis and meta situation. However, only 30% of liver cancer patients are suitable surgery candidates because hepatitis B and hepatitis C worsen the prognosis for cancer meta. Still, surgery is desirable for alleviating the adverse effects of irradiation and so enabling the radiation dose to be increased to 6600rad, thus achieving survival rates equaling those of conventional surgical therapy. Regular treatment is an essential part of liver cancer therapy, and patients must be treated twice daily. Such regular treatment can completely cure over 60% of sufferers. A novel therapeutic method, BID, attacks liver cancer cells with nearly twice the intensity of conventional approaches. Patients treated with BID receive irradiation treatment in the morning to damage the DNA of cancer cells, and a subsequent treatment six to eight hours later further damages the DNA. BID is highly effective in terminating cancer cells but has certain limitations: the liver cancer must be smaller than 5 cm; cancer cells will still be present after surgery; and liver cancer tumors larger than 5-8 cm can not be treated. BID photon knife therapy has high efficacy and reduces damage to normal cells. Integrating BID with other therapeutic methods, including embolism, alcohol injection, surgery or chemotherapy, will ultimately reduce the incidence of liver cancer in Taiwan.

B

How does the advanced PDVC system improve automation control technologies?
By foreseeing potential production bottlenecks

How has the PDVC system demonstrated its success?
By reducing personnel numbers on the factory floor from fifteen to six, while reducing wasted storage capacity from 68,071 to 55,031 units, saving CMC NT$ 2,000,000 over two years

How does the PDVC system cater to individual consumer tastes?
By arranging the simultaneous production of various models, striving to fill factory orders, and making on-line queries regarding how to increase upper and lower supply chain efficiency

C

What have fueled technological advances in manufacturing liquid crystal displays?
Elevated living standards and strong consumer demand in Taiwan

What further demonstrates the phenomenal growth of AU Optronics recently?
Shipments of small- and medium-sized panels increased by 23.4% to reach 3.39 million units during March 2005

What has the ability of the corporation to integrate life sciences and medical technology resulted in?
State-of-the-art photoelectronic laser technology applications

D

Why are there high levels of hepatitis associated liver cancer among Taiwanese?
Because individuals with Hepatitis B have a 150 times greater likelihood of contracting liver cancer than those without the virus

Why are advances in digital technology led to 3D computer-animated control stations widely adopted?
To treat various cancers using radiation therapy

Why is regular treatment an essential part of liver cancer therapy?
Such regular treatment can completely cure over 60% of sufferers.

E

What has ranked as a leading cause of mortality in Taiwan during the past decade?
Liver cancer

How many fatalities does Taiwan have annually from chronic liver hepatitis, cirrhosis or liver cancer?
Over 10,000

How many major therapeutic methods are available for treating liver cancer?

Five

F

Situation 1

1. A 2. C 3. C 4. A 5. B

Situation 2

1. C 2. B 3. A 4. A 5. C

Situation 3

1. C 2. A 3. C 4. B 5. B

G

Situation 1

1. B 2. B 3. C 4. A 5. C 6. B 7. A 8. A 9. C 10. B

Situation 2

1. C 2. A 3. B 4. B 5.C 6. A 7. C 8. B 9. A 10. C 11. C 12. A

Situation 3

1. B 2. B 3. C 4. A 5. A 6. B 7. C 8. B 9. A 10. C 11. C 12. A
13. B 14. C 15. A 16. C 17. C 18. B

I

Situation 4

Many obstacles exist to fully developing e-commerce in Taiwan. Inefficient computerization adopted by many enterprises makes it impossible for enterprises to integrate their efforts with other manufacturers when complying with legal standards and attempting to better understand consumer behavior. Local e-commerce also faces challenges in transaction and distribution. For instance, ensuring network security for online purchases is a key concern. Hackers remain a concern for consumers and enterprises, making customers hesitant to use credit cards online. Technological constraints and cultural considerations mean that the e-commerce sector in Taiwan remains immature. Accelerated development of e-commerce depends on enterprises overcoming barriers and engaging in commercial activities as soon as possible, thus boosting global commerce. Although barriers to market entry are relatively low, returns on investment are high, with potential profits being much higher than in traditional commerce. Additionally, e-commerce can promote brand image in emerging markets. Enterprises must learn to integrate Internet-based services through technical support, marketing strategies, market positioning and collaboration with partners. The rise of the Internet forces enterprises to adopt the above measures to ensure their survival. As the Internet pervades daily life, the workplace and even global commerce, governments and enterprises worldwide must invest heavily in electronic commerce to increase national GDP and corporate competitiveness. To remain abreast of global trends in e-commerce and protect the future standing of Taiwan throughout the Asian-Pacific region, the Taiwanese government drafted and legislated an electronic commerce policy in 1998.

Situation 5

After commercializing the first cellular phone in 1970, ABC Corporation has become the largest global cellular phone manufacturer, diversifying from advanced manufacturing technologies into specialized design processes. With an annual output of 10,000,000 cellular phones, ABC Corporation generated revenues of US$ 1,500,000,000 last year. Manufacturing cellular phones involves integrating electronic equipment, silicon chips, chemical processes, gas water and electricity.

The complexity of cellular phones raises environmental protection, security and hygiene issues in manufacturing. To be a sustainable and environmentally responsible company, ABC Corporation has implemented the following measures: maintaining environmentally safe manufacturing practices aimed at protecting employee health and security; strictly adhering to global safety and hygiene practices; educating employees on the importance of environmental protection, security and hygiene during manufacturing; recommending the adoption of novel environmentally friendly and efficient technologies; and extensively testing new materials to reduce risks. Global enterprises must commit themselves to professional integrity and environmental consciousness. Environmentally sound manufacturing processes should aim to reduce employee error, conserve water and electricity and reduce pollutant emissions. Besides simply generating revenues, ABC Corporation must be a responsible community member by cultivating social consciousness, creating employment opportunities while operating in a sustainable manner, seeking to satisfy consumer demand with quality technological products that have low power consumption and comprehensive technical services.

Situation 6

The Taiwanese Ministry of National Defense has made significant technological advances in manufacturing ground-to-air guided anti-missile systems. For instance, besides developing the Ray-Ting 2000 artillery multiple launch rocket system, the Chung Shan Institute of Science and Technology has designed and enhanced the following missile systems: Shiung-Feng 1 and 2, and Ting-Kung 1 and 2. Previously, Taiwan imported ground-to-air guided anti-missile systems from its diplomatic allies. However, Chinese objections over Taiwanese defense purchases have forced Taiwan to develop its own ground-to-air guided anti-missile system. Shiung-Feng 1 and 2 are short and medium range missile systems that are highly accurate and reliable under adverse weather conditions. Fitted with a semi-active radar homing seeker, the Ting-Kung 1 missile is designed for medium range interception under an aerial saturation attack. The Ting-Kung 2 missile has increased range and firepower by adopting an active radar homing seeker. Additionally, the Ray-Ting 2000, which is designed to repel an amphibious attack and provide superior firepower to conventional tube artillery, is equipped with a fully automatic fire control system, elevation and azimuth driven systems, as well as positional and direction determining systems. This system is considered one of the most powerful artillery multiple launch rocket systems worldwide. To create a balance in the arms buildup involving Taiwan and China, local arms manufacturers such as the Chung Shan Institute of Science and Technology must implement appropriate management strategies to increase their ability to supply domestic defense needs, thus alleviating over dependence on imported weapons.

J

What is a key concern?
Ensuring network security for online purchases

What remains a concern for consumers and enterprises?
Hackers

What does accelerated development of e-commerce depend on?
Enterprises overcoming barriers and engaging in commercial activities as soon as possible

K

How are security and hygiene issues in manufacturing raised?
Owing to the complexity of cellular phones raises environmental protection

How does ABC Corporation hope to be a sustainable and environmentally responsible company?

By maintaining environmentally safe manufacturing practices aimed at protecting employee health and security; strictly adhering to global safety and hygiene practices; educating employees on the importance of environmental protection, security and hygiene during manufacturing; recommending the adoption of novel environmentally friendly and efficient technologies; and extensively testing new materials to reduce risks.

How does ABC Corporation intend to be a responsible community member?

By cultivating social consciousness, creating employment opportunities while operating in a sustainable manner, seeking to satisfy consumer demand with quality technological products that have low power consumption and comprehensive technical services

L

Why has the Taiwanese Ministry of National Defense made significant technological advances?

Owing to its ability to manufacture ground-to-air guided anti-missile systems

Why is Taiwan forced to develop its own ground-to-air guided anti-missile system?

Because of Chinese objections over Taiwanese defense purchases

Why must local arms manufacturers such as the Chung Shan Institute of Science and Technology implement appropriate management strategies to increase their ability to supply domestic defense needs?

To alleviate over dependence on imported weapons

M

What has the Taiwanese Ministry of National Defense made significant technological advances in?
Manufacturing ground-to-air guided anti-missile systems

What has the Chung Shan Institute of Science and Technology has designed and enhanced?
Shiung-Feng 1 and 2

What is designed to repel an amphibious attack and provide superior firepower to conventional tube artillery?
The Ray-Ting 2000

N

Situation 4

1. A 2. C 3. B 4. B 5. C

Situation 5

1. B 2. C 3. A 4. A 5. C

Situation 6

1. B 2. C 3. B 4. A 5. C

O

Situation 4

1. C 2. C 3. B 4. B 5. A 6. C 7. A 8. C 9. A 10. C 11. B 12. C
13. B

Situation 5

1. A 2. C 3. B 4. C 5. C 6. B 7. A 8. A 9. B 10. A

Situation 6

1. C 2. A 3. C 4. C 5. B 6. B 7. C 8. A 9. C 10. B 11. A

A

Situation 1

The Taiwanese animation sector is involved in animation films, cartoons, animation videos and Internet-based animation. Digitization, or placing animated objects in a digital context, has transformed the sector and increased productivity and innovativeness. Still, the animation sector faces many obstacles. Overseas businesses continue to view Taiwanese animation firms as original equipment manufacturers (OEMs), limiting their ability to build brand recognition. Additionally, labor costs in China and India are significantly lower than in Taiwan. Moreover, Taiwanese animation companies lack animation designers with global marketing expertise. Taiwanese animation companies also have difficulty in securing bank loans or other financial support as company startups, largely owing to their relative immaturity. Despite these obstacles, the local animation sector is increasingly pivotal in the economic development of Taiwan, and several encouraging developments are underway. For example, the Ministry of Economic Affairs recently initiated a loan fund to develop digital content, which will benefit the animation sector. In addition to providing the animation sector a more solid economic foundation, this project will also encourage an influx of skilled technical and management personnel. Additionally, the Taiwanese animation sector has adopted increasingly novel 3D techniques in cartoons, animation videos and Internet-based animation and animation films, for example Toy Story and SHREK. Organizations in Taiwan committed to developing animation-related techniques can be categorized as either academic research centers or R&D departments in animation firms, with the former focused on developing 3D techniques and animation designs and the latter focused on business-related research and development tasks, such as enhancing process efficiency. If the Taiwanese animation sector is to become globally competitive, it must move beyond the traditional OEM model and develop brand recognition.

Situation 2

Taiwan officially became an aging society in 1997 according to the definition of the United Nations. To look after the welfare of the elderly, the government needs to provide affordable, basic housing that suits the needs of this growth sector. Local construction enterprises thus have begun promoting retirement housing complexes, which can be categorized into four types. First, luxury accommodations are available for retirees, which have restrictions on tenant age and health status. Leased to tenants for long periods ranging from three to twenty years, such communities are normally referred to as congregate housing, and are characterized by uniform housing, organized social activities, health management, hotel type services, and specialist service staff. Such housing arrangements offer a diverse array of services to high income elderly individuals. Second, government-registered residential facilities are available for the elderly, which are subsidized by the government with elderly residents contributing the remainder themselves. Such facilities provide only basic daily services and so are relatively inexpensive. Third, government-registered residential facilities are run in conjunction with private recuperative centers. These facilities mainly house bed-ridden elderly, and are staffed by general nursing staff without specialized skills and who simply provide primary care. Fourth, long-term residential care facilities are available for chronically ill elderly who are largely bed ridden. These facilities differ from long-term care facilities in hospitals in that they aim mainly to support the daily functions of the elderly, while hospitals focus on treating chronic illnesses. Notable examples of these facilities include residential facilities for retired serviceman that combine housing with recuperative facilities. What all of the above facilities share in common is that they are residential nursing homes staffed with professional nurses.

Situation 3

Taiwan is a key global manufacturer of 3C and synthetic fiber products. The current trend towards using nano-materials in fiber production creates enormous potential for further development of this industry. The numerous existing nano-material applications for locally produced 3C and synthetic fiber products include inorganic nano-recorder media, nano-interface handle for a separate membrane in batteries and electron passive element. The ability to adopt advanced production technologies such as nano-materials is important to Taiwan's continued economic growth. Taiwanese manufacturers must remain abreast of the latest applications of nano-elements and organisms, while also considering environmental sustainability and using energy efficiently for manufacturing purposes. Since the Taiwanese economy has matured to the extent where labor and capital resources are already efficiently allocated for manufacturing, continued economic development depends on the ability to constantly adopt advanced manufacturing practices such as nano-materials technology. Accordingly, the Taiwanese government has offered numerous incentives for local companies to develop their research capabilities, and these incentives have seen Taiwan transform from an agricultural-based economy to a hi-tech one over just a few decades. During the coming decade, Taiwan faces the following challenges in further developing its nano-materials technology sector: developing and synthesizing nano-materials efficiently, manipulating the properties of nano-materials to meet industrial specifications, and further understanding the properties of nano-materials and their implications for product commercialization. One particular concern is how to reduce the time between product development and commercialization. Taiwan must effectively meet the above challenges to remain competitive in the fiercely competitive nanotechnology sector.

B

Why do overseas businesses continue to view Taiwanese animation firms as original equipment manufacturers (OEMs)?
Owing to their inability to build brand recognition

Why has the Ministry of Economic Affairs recently initiated a loan fund?
To develop digital content, which will benefit the animation sector

Why must the Taiwanese animation sector move beyond the traditional OEM model and develop brand recognition?
To become globally competitive

C

What have restrictions on tenant age and health status?
Luxury accommodations for retirees

What are subsidized by the government with elderly residents contributing the remainder themselves?
Government-registered residential facilities

What do all of the above facilities share in common?
They are residential nursing homes staffed with professional nurses

D

Which trend creates enormous potential for further development of 3C and synthetic fiber products?
The one using nano-materials in fiber production

Which applications for locally produced 3C and synthetic fiber products include inorganic nano-recorder media, nano-interface handle for a separate membrane in batteries and electron passive elements?
Nano-material ones

Which country is a key global manufacturer of 3C and synthetic fiber products?

Taiwan

Which factor is important to Taiwan's continued economic growth?

The ability to adopt advanced production technologies such as nano-materials

Which incentives have seen Taiwan transform from an agricultural-based economy to a hi-tech one over just a few decades?

Those for local companies to develop their research capabilities

E

What are among the numerous existing nano-material applications for locally produced 3C and synthetic fiber products?

Inorganic nano-recorder media, nano-interface handle for a separate membrane in batteries and electron passive elements

What has the Taiwanese government offered?

Numerous incentives for local companies to develop their research capabilities

What is important to Taiwan's continued economic growth?

The ability to adopt advanced production technologies such as nano-materials

F

Situation 1

1. B 2. A 3. C 4. B 5. B

Situation 2

1. C 2. B 3. C 4. A 5. C

Situation 3

1. A 2. C 3. B 4. A 5. C

G

Situation 1

1. C 2. C 3. A 4. A 5. C 6. B 7. B 8. C 9. A 10. C 11. B 12. B
13. C

Situation 2

1. B 2. B 3. C 4. A 5. C 6. A 7. B 8. A 9. C 10. C 11. B 12. C
13. B 14. B

Situation 3

1. C 2. B 3. C 4. A 5. C 6. C 7. B 8. C 9. A 10. C

I

Situation 4

The Taiwan Rail Administration (TRA) oversees conventional rail transportation in
Taiwan. Since Liu Ming-Chuan directed the construction of the first rail line from
Keelung to Taipei in 1886, rail has been important to long-distance transportation in
Taiwan. In response to rapid population and economic growth, TRA completed the
north-link line, the double link of the western line, the electrified project of the
western line, the purchase of the push-pull Tzu-Chiang express train and, most
significantly, the south-link line and the round island railway network. Recent
passenger numbers have not justified the tremendous capital requirements of railway
construction. Notably, completion of the Chung-shan and Formosa freeway projects
encouraged individuals to drive cars or take passenger buses instead of riding the
train, and many bus companies offer an attractive combination of luxurious seating

and inexpensive off-peak ticket prices. To encourage passengers to return to railway transportation, TRA has promoted the Taipei-Kaohsiung Tzu-Chiang express train (3 hours 59 minutes one way) and the Taipei-Hualien Chu-Kuang group express train. The Taiwan High Speed Railway (THSR) has been the most daunting undertaking of TRA to date. To realize the THSR, TRA is directing several construction projects, including commuter stations in San-Keng, Tai-Yuan, Da-Ching and Da-Chiao, as well as a metro hub linking Taipei and Hsinchu. To make the THSR competitive with other transportation modes, TRA needs to provide additional short-distance train lines, promote custom-designed train tourism packages, provide shuttle services linking THSR stations with downtown areas and expand train cargo capacity. Moreover, the Taiwan Department of Transportation should integrate the efforts of TRA and the THSR to provide a comprehensive, comfortable and convenient passenger service.

Situation 5

During its occupation of Taiwan, Japan sought to cultivate local medical talent by discouraging its own physicians from practicing in Taiwan. Moreover, the Japanese established a medical training institute in 1897 at a teaching hospital facility in Taipei, which in 1899 became a medical school with its own specialized departments of medical science. However, this facility catered mainly to Japanese governmental officials or the privileged gentry class in Taipei rather than catering to the general public. Later, teaching hospitals were established, which aimed to serve local community needs. Eventually, the Japanese Red Cross Society in Taiwan was established in 1904. Moreover, the Taipei University of the Japanese Prefecture set up a medical department in 1935, some years after the university's original establishment in 1928. After Chiang-Kai Shek and his Kuomingtang Government arrived in Taiwan in the wake of the civil war, problems such as inflationary prices and scarcity of medicine led to the emergence of unlicensed physicians and poor quality medicines. However, since 1965, medical practices and standards in Taiwan have gradually improved to reach the current situation where Taiwan provides state-of-the-art medical services. Taiwanese hospitals are categorized either as medical centers (17), regional hospitals (70), community hospitals (375) or clinics (data unavailable). Since its establishment in 1995, the National Health Insurance scheme has sought to provide medical coverage for all Taiwanese residents under the auspices of the National Health Insurance Bureau. However, the National Health Insurance scheme suffers from a severe fiscal imbalance, causing near bankruptcy and other problems for medical facilities that rely on NHI subsidies. Consequently, a large island-wide protest march of medical professionals was held this year.

Situation 6

Substandard corporate performances have driven down the prices of many listed stocks in Taiwan, straining the Taiwanese financial sector by increasing loan default rates. To strengthen the financial sector, the Taiwanese government has initiated comprehensive reforms focused on market liberalization and institutional mergers. However, these well-intentioned policies have incurred further financial problems. For example, market liberalization policies have led to the emergence of large numbers of banks, saturating the market to the point where banking institutions suffer low profits and have difficulty differentiating themselves from competitors. Additionally, institutional mergers failed to resolve basic financial problems. The merger policies were intended to strengthen poorly performing banks or credit cooperatives by merging them with more healthy financial institutions. However, merged entities have suffered impaired corporate performance owing to defaulted loans and debts acquired during their mergers. With continued globalization, most notably in the form of WTO entry, the financial sector will undoubtedly face increased competition as overseas firms gain a foothold in the domestic market. Furthermore, given the financial problems associated with market liberalization and institutional merger policies, the Taiwanese government must closely re-examine the problems and implement responses to increase banking service quality and improve the competitiveness of domestic institutions.

J

What encouraged individuals to drive cars or take passenger buses instead of riding the train?
Completion of the Chung-shan and Formosa freeway projects

What has TRA done to encourage passengers to return to railway transportation?
It has promoted the Taipei-Kaohsiung Tzu-Chiang express train (3 hours 59 minutes one way) and the Taipei-Hualien Chu-Kuang group express train.

What must TRA do to make the THSR competitive with other transportation modes?
It needs to provide additional short-distance train lines, promote custom-designed train tourism packages, provide shuttle services linking THSR stations with downtown areas and expand train cargo capacity.

K

When did the teaching hospital facility in Taipei become a medical school with its own specialized departments of medical science?
In 1899

When was the Japanese Red Cross Society in Taiwan established?
In 1904

When did the Taipei University of the Japanese Prefecture set up a medical department?
In 1935

L

Why has the Taiwanese financial sector become strained?

Because substandard corporate performances have driven down the prices of many listed stocks in Taiwan

Why has the market become saturated to the point where banking institutions suffer low profits and have difficulty differentiating themselves from competitors?

Because market liberalization policies have led to the emergence of large numbers of banks

Why must the Taiwanese government closely re-examine the problems and implement responses to increase banking service quality and improve the competitiveness of domestic institutions?

Because of the financial problems associated with market liberalization and institutional merger policies

M

How has the Taiwanese Government attempted to strengthen the financial sector?

By initiating comprehensive reforms focused on market liberalization and institutional mergers

What have the merger policies intended to do?

Strengthen poorly performing banks or credit cooperatives by merging them with more healthy financial institutions

Why will the financial sector in Taiwan undoubtedly face increased competition?

As overseas firms gain a foothold in the domestic market

N

Situation 4

1. A 2. C 3. C 4. B 5. A

Situation 5

1. B 2. A 3. B 4. A 5. C

Situation 6

1. C 2. A 3. C 4. B 5. A

O

Situation 4

1. B 2. C 3. A 4. C 5. A 6. C 7. B 8. B 9. C 10. A

Situation 5

1. B 2. A 3. A 4. C 5. C 6. A 7. C 8. B 9. B 10. A 11. C

Situation 6

1. B 2. C 3. A 4. C 5. A 6. B 7. A 8. C 9. B 10. C

About the Author

Born on his father's birthday, Ted Knoy received a Bachelor of Arts in History at Franklin College of Indiana (Franklin, Indiana) and a Master's degree in Public Administration at American International College (Springfield, Massachusetts). He is currently a Ph.D. student in Education at the University of East Anglia (Norwich, England). Having conducted research and independent study in New Zealand, Ukraine, Scotland, South Africa, India, Nicaragua and Switzerland, he has lived in Taiwan since 1989 where he has been a permanent resident since 2000.

Having taught technical writing in the graduate school programs of National Chiao Tung University (Institute of Information Management, Institute of Communications Engineering and, currently, in the College of Management) and National Tsing Hua University (Computer Science, Life Science, Electrical Engineering, Power Mechanical Engineering, Chemistry and Chemical Engineering Departments) since 1989, Ted also teaches in the Institute of Business Management at Yuan Pei University of Science and Technology. He is also the English editor of several technical and medical journals and publications in Taiwan.

Ted is author of *The Chinese Technical Writers' Series,* which includes An English Style Approach for Chinese Technical Writers, English Oral Presentations for Chinese Technical Writers, A Correspondence Manual for Chinese Technical Writers, An Editing Workbook for Chinese Technical Writers and Advanced Copyediting Practice for Chinese Technical Writers. He is also author of *The Chinese Professional Writers' Series,* which includes Writing Effective Study Plans, Writing Effective Work Proposals, Writing Effective Employment Application Statements, Writing Effective Career Statements, Effectively Communicating Online and Writing Effective Marketing Promotional Materials.

Ted created and coordinates the Chinese On-line Writing Lab (OWL) at www.chineseowl.idv.tw

Acknowledgments

Thanks to the following individuals for contributing to this book:

元培科學技術學院 經營管理研究所

許碧芳（所長）	王貞穎	李仁智	陳彥谷	胡惠眞	陳碧俞	王連慶
蔡玟純	高青莉	賴姝惠	李雅玎	戴碧美	楊明雄	陳皇助	林宏隆
鍾玽融	李昭蓉	許美菁	葉伯彥	林羿君	吳政龍	鄭凱元	黃志斌
郭美萱	李尉誠	陳靜怡	盧筱嵐	鄭彥均	劉偉翔	彭廣興	林宗瑋
巫怡樺	朱建華					

元培科學技術學院 影像醫學研究所

王愛義（所長）	周美榮	顏映君	林孟聰	張雅玲	彭薇莉	張明偉
李玉綸	聶伊辛	黃勝賢	張格瑜	龔慧敏	林永健	呂忠祐	李仁忠
王國偉	李政翰	黃國明	蔡明輝	杜俊元	丁健益	方詩涵	余宗銘
劉力瑛	郭明杰					

元培科學技術學院 生物技術研究所

陳媛孃（所長）	范齡文	彭姵華	鄭啓軒	許凱文	李昇憲	陳雪君
鄭凱暹	尤鼎元	陳玉梅	鄭美玲	郭軒中	朱芳儀	周佩穎	吳佳眞

Thanks also to Ming-Jay Chen for illustrating this book and Yang Ming-Hsiung for providing technical support. Graduate students at Yuanpei University of Science and Technology in the Institute of Business Management, the Institute of Biotechnology and the Institute of Medical Imagery are also appreciated. My technical writing students in the Department of Computer Science and Institute of Life Science at National Tsing Hua University, as well as the College of Management at National Chiao Tung University are also appreciated. Thanks also to Seamus Harris and Bill Johnson for reviewing this workbook.

精通科技論文（報告）寫作之捷徑
An English Style Approach for Chinese Technical Writers （修訂版）

作者：柯泰德（Ted Knoy）

內容簡介
使用直接而流利的英文會話
讓您所寫的英文科技論文很容易被了解
提供不同形式的句型供您參考利用
比較中英句子結構之異同
利用介系詞片語將二個句子連接在一起

萬其超 / 李國鼎科技發展基金會秘書長

本書是多年實務經驗和專注力之結晶，因此是一本坊間少見而極具實用
價值的書。

陳文華 / 國立清華大學工學院院長

中國人使用英文寫作時，語法上常會犯錯，本書提供了很好的實例示
範，對於科技論文寫作有相當參考價值。

徐　章 / 工業技術研究院量測中心主任

這是一個讓初學英文寫作的人，能夠先由不犯寫作的錯誤開始再根據書
中的步驟逐步學習提升寫作能力的好工具，此書的內容及解說方式使讀
者也可以無師自通，藉由自修的方式學習進步，但是更重要的是它雖然
是一本好書，當您學會了書中的許多技巧，如果您還想要更進步，那麼
基本原則還是要常常練習，才能發揮書的精髓。

**Kathleen Ford, English Editor, Proceedings(Life Science Divison),
National Science Council**

The Chinese Technical Writers Series is valuable for anyone involved with
creating scientific documentation.

※若有任何英文文件修改問題，請直接與柯泰德先生聯絡： (03) 5724895

特　　價　新台幣300元
劃　　撥　19419482 清蔚科技股份有限公司
線上訂購　四方書網 www.4Book.com.tw
發 行 所　華香園出版社

作好英語會議簡報
English Oral Presentations for Chinese Technical Writers

作者：柯泰德（Ted Kony）

內容簡介

本書共分十二個單元，涵括產品開發、組織、部門、科技、及產業的介紹、科技背景、公司訪問、研究能力及論文之發表等，每一單元提供不同型態的科技口頭簡報範例，以進行英文口頭簡報的寫作及表達練習，是一本非常實用的著作。

李鍾熙 / 工業技術研究院化學工業研究所所長

　　一個成功的科技簡報，就是使演講流暢，用簡單直接的方法、清楚表達內容。本書提供一個創新的方法（途徑），給組織每一成員做為借鏡，得以自行準備口頭簡報。利用本書這套有系統的方法加以練習，將必然使您信心備增，簡報更加順利成功。

薛敬和 / IUPAC台北國際高分子研討會執行長
　　　　　國立清華大學教授

　　本書以個案方式介紹各英文會議簡報之執行方式，深入簡出，為邁入實用狀況的最佳參考書籍。

沙晉康 / 清華大學化學研究所所長
　　　　　第十五屆國際雜環化學會議主席

　　本書介紹英文簡報的格式，值得國人參考。今天在學術或工商界與外國接觸來往均日益增多，我們應加強表達的技巧，尤其是英文的簡報應具有很高的專業水準。本書做為一個很好的範例。

張俊彥 / 國立交通大學電機資訊學院教授兼院長

　　針對中國學生協助他們寫好英文的國際論文參加國際會議如何以英語演講、內容切中要害特別推薦。

※若有任何英文文件修改問題，請直接與柯泰德先生聯絡：（03）5724895

特　　　價　　新台幣250元
劃　　　撥　　19419482 清蔚科技股份有限公司
線上訂購　　四方書網 www.4Book.com.tw
發 行 所　　工業技術研究院

英文信函參考手冊
A Correspondence Manual for Chinese Technical Writers

作者：柯泰德（Ted Knoy）

內容簡介

本書期望成為從事專業管理與科技之中國人，在國際場合上溝通交流時之參考指導書籍。本書所提供的書信範例（附磁碟片），可為您撰述信件時的參考範本。更實際的是，本書如同一「寫作計畫小組」，能因應特定場合（狀況）撰寫出所需要的信函。

李國鼎／總統府資政

我國科技人員在國際場合溝通表達之機會急遽增加，希望大家都來重視英文說寫之能力。

羅明哲／國立中興大學教務長

一份表達精準且適切的英文信函，在國際間的往來交流上，重要性不亞於研究成果的報告發表。本書介紹各類英文技術信函的特徵及寫作指引，所附範例中肯實用，為優良的學習及參考書籍。

廖俊臣／國立清華大學理學院院長

本書提供許多有關工業技術合作、技術轉移、工業資訊、人員訓練及互訪等接洽信函的例句和範例，頗為實用，極具參考價值。

于樹偉／工業安全衛生技術發展中心主任

國際間往來日益頻繁，以英文有效地溝通交流，是現今從事科技研究人員所需具備的重要技能。本書在寫作風格、文法結構與取材等方面，提供極佳的寫作參考與指引，所列舉的範例，皆經過作者細心的修訂與潤飾，必能切合讀者的實際需要。

※若有任何英文文件修改問題，請直接與柯泰德先生聯絡：（03）5724895

特　　價　新台幣250元
劃　　撥　19419482 清蔚科技股份有限公司
線上訂購　四方書網 www.4Book.com.tw
發 行 所　工業技術研究院

科技英文編修訓練手冊
An Editing Workbook for Chinese Technical Writers

作者：柯泰德（Ted Knoy）

內容簡介

要把科技英文寫的精確並不是件容易的事情。通常在投寄文稿發表前，作者都要前前後後修改草稿，在這樣繁複過程中甚至最後可能請專業的文件編修人士代勞雕琢使全文更為清楚明確。

本書由科技論文的寫作型式、方法型式、內容結構及內容品質著手，並以習題方式使學生透過反覆練習熟能生巧，能確實提昇科技英文之寫作及編修能力。

劉炯明／國立清華大學校長

「科技英文寫作」是一項非常重要的技巧。本書針對台灣科技研究人員在英文寫作發表這方面的訓練，書中以實用性練習對症下藥，期望科技英文寫作者熟能生巧，實在是一個很有用的教材。

彭旭明／國立台灣大學副校長

本書為科技英文寫作系列之四；以練習題為主，由反覆練習中提昇寫作反編輯能力。適合理、工、醫、農的學生及研究人員使用，特為推薦。

許千樹／國立交通大學研究發展處研發長

處於今日高科技時代，國人用到科技英文寫作之機會甚多，如何能以精練的手法寫出一篇好的科技論文，極為重要。本書針對國人寫作之缺點提供了各種清楚的編修範例，實用性高，極具參考價值。

陳文村／國立清華大學電機資訊學院院長

處在我國日益國際化、資訊化的社會裡，英文書寫是必備的能力，本書提供很多極具參考價值的範例。柯泰德先生在清大任教科技英文寫作多年，深受學生喜愛，本人樂於推薦此書。

※若有任何英文文件修改問題，請直接與柯泰德先生聯絡：（03）5724895

特　　　價　新台幣350元
劃　　　撥　19419482 清蔚科技股份有限公司
線上訂購　四方書網 www.4Book.com.tw
發 行 所　清蔚科技股份有限公司

科技英文編修訓練手冊【進階篇】
Advanced Copyediting Practice for Chinese Technical Writers

作者：柯泰德（Ted Knoy）

內容簡介

本書延續科技英文寫作系列之四「科技英文編修訓練手冊」之寫作指導原則，更進一步把重點放在如何讓作者想表達的意思更明顯，即明白寫作。把文章中曖昧不清全部去除，使閱讀您文章的讀者很容易的理解您作品的精髓。

本手冊同時國立清華大學資訊工程學系非同步遠距教學科技英文寫作課程指導範本。

張俊彥 / 國立交通大學校長暨中研院院士

對於國內理工學生及從事科技研究之人士而言，可說是一本相當有用的書籍，特向讀者推薦。

蔡仁堅 / 前新竹市長

科技不分國界，隨著進入公元兩千年的資訊時代，使用國際語言撰寫學術報告已是時勢所趨；今欣見柯泰德先生致力於編撰此著作，並彙集了許多實例詳加解說，相信對於科技英文的撰寫有著莫大的裨益，特予以推薦。

史欽泰 / 工研院院長

本書即以實用範例，針對國人寫作的缺點提供簡單、明白的寫作原則，非常適合科技研發人員使用。

張智星 / 國立清華大學資訊工程學系副教授、計算中心組長

本書是特別針對系上所開科技英文寫作非同步遠距教學而設計，範圍內容豐富，所列練習也非常實用，學生可以配合課程來使用，在時間上更有彈性的針對自己情況來練習，很有助益。

劉世東 / 長庚大學醫學院微生物免疫科主任

書中的例子及習題對閱讀者會有很大的助益。這是一本研究生必讀的書，也是一般研究者重要的參考書。

※若有任何英文文件修改問題，請直接與柯泰德先生聯絡：（03）5724895

特　　價　新台幣450元
劃　　撥　19419482 清蔚科技股份有限公司
線上訂購　四方書網 www.4Book.com.tw
發 行 所　清蔚科技股份有限公司

有效撰寫英文讀書計畫
Writing Effective Study Plans

作者：柯泰德（Ted Knoy）

內容簡介

本書指導準備出國進修的學生撰寫精簡切要的英文讀書計畫，內容包括：表達學習的領域及興趣、展現所具備之專業領域知識、敘述學歷背景及成就等。本書的每個單元皆提供視覺化的具體情境及相關寫作訓練，讓讀者進行實際的訊息運用練習。此外，書中的編修訓練並可加強「精確寫作」及「明白寫作」的技巧。本書適用於個人自修以及團體授課，能確實引導讀者寫出精簡而有效的英文讀書計畫。

本手冊同時為國立清華大學資訊工程學系非同步遠距教學科技英文寫作課程指導範本。

于樹偉／工業技術研究院主任

> 《有效撰寫讀書計畫》一書主旨在提供國人精深學習前的準備，包括：讀書計畫及推薦信函的建構、完成。藉由本書中視覺化訊息的互動及練習，國人可以更明確的掌握全篇的意涵，及更完整的表達心中的意念。這也是本書異於坊間同類書籍只著重在片斷記憶，不求理解最大之處。

王　玫／工業研究技術院、化學工業研究所組長

> 《有效撰寫讀書計畫》主要是針對想要進階學習的讀者，由基本的自我學習經驗描述延伸至未來目標的設定，更進一步強調推薦信函的撰寫，藉由圖片式訊息互動，讓讀者主動聯想及運用寫作知識及技巧，避免一味的記憶零星的範例；如此一來，讀者可以更清楚表明個別的特質及快速掌握重點。

※若有任何英文文件修改問題，請直接與柯泰德先生聯絡：（03）5724895

特　　價　新台幣450元
劃　　撥　19419482 清蔚科技股份有限公司
線上訂購　四方書網 www.4Book.com.tw
發 行 所　清蔚科技股份有限公司

有效撰寫英文工作提案
Writing Effective Work Proposals

作者：柯泰德（Ted Knoy）

內容簡介

許多國人都是在工作方案完成時才開始撰寫相關英文提案，他們視撰寫提案為行政工作的一環，只是消極記錄已完成的事項，而不是積極的規劃掌控未來及現在正進行的工作。如果國人可以在撰寫英文提案時，事先仔細明辨工作計畫提案的背景及目標，不僅可以確保寫作進度、寫作結構的完整性，更可兼顧提案相關讀者的興趣強調。本書中詳細的步驟可指導工作提案寫作者達成此一目標。 書中的每個單元呈現三個視覺化的情境，提供國人英文工作提案寫作實質訊息，而相關附加的寫作練習讓讀者做實際的訊息運用。此外，本書也非常適合在課堂上使用，教師可以先描述單元情境而讓學生藉由書中練習循序完成具有良好架構的工作提案。書中內容包括：1.工作提案計畫（第一部分）：背景 2.工作提案計畫（第二部分）：行動 3.問題描述 4.假設描述 5.摘要撰寫（第一部分）： 簡介背景、目標及方法 6.摘要撰寫（第二部分）： 歸納希望的結果及其對特定領域的貢獻 7.綜合上述寫成精確工作提案。

唐傳義／國立清華大學資訊工程學系主任

本書重點放在如何在工作計畫一開始時便可以用英文來規劃整個工作提案，由工作提案的背景、行動、方法及預期的結果漸次教導國人如何寫出具有良好結構的英文工作提案。如此用英文明確界定工作提案的程序及工作目標更可以確保英文工作提案及工作計畫的即時完成。對工作效率而言也有助益。

在國人積極加入WTO之後的調整期，優良的英文工作提案寫作能力絕對是一項競爭力快速加分的工具。

※若有任何英文文件修改問題，請直接與柯泰德先生聯絡：（03）5724895

特　　價　　新台幣450元
劃　　撥　　19735365 葉忠賢
線上訂購　　www.ycrc.com.tw
發 行 所　　揚智文化事業股份有限公司

有效撰寫求職英文自傳
Writing Effective Employment Application Statements

作者：柯泰德（Ted Knoy）

內容簡介

本書主要教導讀者如何建構良好的求職英文自傳。書中內容包括：1.表達工作相關興趣；2.興趣相關產業描寫；3.描述所參與方案裡專業興趣的表現；4.描述學歷背景及已獲成就；5.介紹研究及工作經驗；6.描述與求職相關的課外活動；7.綜合上述寫成精確求職英文自傳。

有效的求職英文自傳不僅必須能讓求職者在企業主限定的字數內精確的描述自身的背景資訊及先前成就，更關鍵性的因素是有效的求職英文自傳更能讓企業主快速明瞭求職者如何應用相關知識技能或其特殊領導特質來貢獻企業主。

書中的每個單元呈現三個視覺化的情境，提供國人求職英文自傳寫作實質訊息，而相關附加的寫作練習讓讀者做實際的訊息運用。此外，本書也非常適合在課堂上使用，教師可以先描述單元情境而讓學生藉由書中練習循序完成具有良好架構的求職英文自傳。

黎漢林／國立交通大學管理學院院長

我國加入WTO後，國際化的腳步日益加快；而企業人員之英文寫作能力更形重要。它不僅可促進國際合作夥伴間的溝通，同時也增加了國際客戶的信任。因此國際企業在求才時無不特別注意其員工的英文表達能力。

柯泰德先生著作《有效撰寫求職英文自傳》即希望幫助求職者能以英文有系統的介紹其能力、經驗與抱負。這本書是柯先生有關英文寫作的第八本專書，柯先生教學與編書十分專注，我相信這本書對求職者是甚佳的參考書籍。

※若有任何英文文件修改問題，請直接與柯泰德先生聯絡：（03）5724895

特　　價　新台幣450元
劃　　撥　19735365 葉忠賢
線上訂購　www.ycrc.com.tw
發 行 所　揚智文化事業股份有限公司

有效撰寫英文職涯經歷
Writing Effective Career Statements

作者：柯泰德（Ted Knoy）

內容簡介

本書主要教導讀者如何建構良好的英文職涯經歷。書中內容包括：1.表達工作相關興趣；2.興趣相關產業描寫；3.描述所參與方案裡專業興趣的表現；4.描述學歷背景及已獲成就；5.介紹研究及工作經驗；6.描述與求職相關的課外活動；7.綜合上述寫成英文職涯經歷。

有效的職涯經歷描述不僅能讓再度就業者在企業主限定的字數內精準的描述自身的背景資訊及先前工作經驗及成就，更關鍵性的，有效的職涯經歷能讓企業主快速明瞭求職者如何應用相關知識技能及先前的就業經驗結合來貢獻企業主。

書中的每個單元呈現六個視覺化的情境，經由以全民英語檢定為標準而設計的口說訓練、聽力、閱讀及寫作四種不同功能來強化英文能力。此外，本書也非常適合在課堂上使用，教師可以先描述單元情境而讓學生藉由書中練習循序在短期內完成。

林進財／元培科學技術學院校長

近年來，台灣無不時時刻刻地努力提高國際競爭力，不論政府或企業界求才皆以英文表達能力為主要考量之一。唯有員工具備優秀的英文能力，才足以把本身的能力、工作經驗與國際競爭舞台接軌。

柯泰德先生著作《有效撰寫英文職涯經歷》，即希望幫助已有工作經驗的求職者能以英文有效地介紹其能力、工作經驗與成就。此書是柯先生有關英文寫作的第九本專書，相信對再度求職者是進入職場絕佳的工具書。

※若有任何英文文件修改問題，請直接與柯泰德先生聯絡：（03）5724895

特　　　價　新台幣480元
劃　　　撥　19735365 葉忠賢
線上訂購　www.ycrc.com.tw
發 行 所　揚智文化事業股份有限公司

有效撰寫專業英文電子郵件
Effectively Communicating Online

作者：柯泰德（Ted Knoy）

內容簡介

本書主要教導讀者如何建構良好的專業英文電子郵件。書中內容包括：1.科技訓練請求信函；2.資訊交流信函；3.科技訪問信函；4.演講者邀請信函；5.旅行安排信函；6.資訊請求信函。

書中的每個單元呈現三個視覺化的情境，經由以全民英語檢定為標準而設計的口說訓練、聽力、閱讀及寫作四種不同功能來強化英文能力。此外，本書也非常適合在課堂上使用，教師可以先描述單元情境而讓學生藉由書中練習循序在短期內完成。

許碧芳／元培科學技術學院經營管理研究所所長

隨著時代快速變遷，人們生活步調及習性也十倍速的演變。舉郵件為例，由早期傳統的郵局寄送方式改為現今的電子郵件（e-mail）系統。速度不但快且也節省費用。對有時效性的訊息傳送更可達事半功倍的效果。不僅如此，電子郵件不受地域性的限制，可以隨地進行溝通，也是生活及職場上一項利器。

柯先生所著《有效撰寫專業英文電子郵件》，乃針對目前對電子郵件寫作需求，配合六種不同情境展示近二百個範例寫作。藉此觀摩他人電子郵件寫作來加強讀者本身的寫作技巧，同時配合書中網路練習訓練英文聽力及閱讀技巧。是一本非常實用且符合網路時代需求的工具書。

※若有任何英文文件修改問題，請直接與柯泰德先生聯絡：（03）5724895

特　　價　新台幣520元
劃　　撥　19735365 葉忠賢
線上訂購　www.ycrc.com.tw
發 行 所　揚智文化事業股份有限公司

The Chinese
Online Writing Lab

【 柯泰德線上英文論文編修訓練服務 】

http://www.cc.nctu.edu.tw/~tedknoy

您有科技英文寫作上的困擾嗎？

您的文章在投稿時常被國外論文審核人員批評文法很爛嗎？以至於被退稿嗎？

您對論文段落的時式使用上常混淆不清嗎？

您在寫作論文時同一個動詞或名詞常常重複使用嗎？

您的這些煩惱現在均可透過柯泰德網路線上科技英文論文編修服務來替您加以解決。本服務項目分別含括如下：

1. 英文論文編輯與修改
2. 科技英文寫作開課訓練服務
3. 線上寫作家教
4. 免費寫作格式建議服務，及網頁問題討論區解答
5. 線上遠距教學（互動練習）

另外，為能廣為服務中國人士對論文寫作上之缺點，柯泰德亦同時著作下列參考書籍可供有志人士為寫作上之參考。

＜1.精通科技論文（報告）寫作之捷徑
＜2.做好英文會議簡報
＜3.英文信函參考手冊
＜4.科技英文編修訓練手冊
＜5.科技英文編修訓練手冊（進階篇）
＜6.有效撰寫英文讀書計畫

上部分亦可由柯泰德先生的首頁中下載得到。

如果您對本服務有興趣的話，可參考柯泰德先生的首頁標示。

柯泰德網路線上科技英文論文編修服務
地址：新竹市大學路50號8樓之三
TEL:03-5724895
FAX:03-5724938
網址：http://www.cc.nctu.edu.tw/~tedknoy
E-mail:tedaknoy@ms11.hinet.net

應用英文寫作系列 06

有效撰寫行銷英文

作　　者／柯泰德（Ted Knoy）
出 版 者／揚智文化事業股份有限公司
發 行 人／葉忠賢
總 編 輯／閻富萍
執行編輯／胡琡珮
地　　址／台北縣深坑鄉北深路三段 260 號 8 樓
電　　話／(02)2664-7780
傳　　真／(02)2664-7633
　E-mail ／ service@ycrc.com.tw
印　　刷／鼎易印刷事業股份有限公司
　I S B N ／ 978-957-818-836-5
初版一刷／2007 年 9 月
定　　價／新台幣 480 元

國家圖書館出版品預行編目資料

有效撰寫行銷英文 / 柯泰德 (Ted Knoy) 作.
-- 初版. -- 臺北縣深坑鄉：揚智文化，
2007.09
  面；　公分. --（應用英文寫作系列；
6）

ISBN 978-957-818-836-5（平裝）

1.商業英文　2.商業應用文

493.6　　　　　　　　　　96016422